TRAFFIC THEORY

INTERNATIONAL SERIES IN OPERATIONS RESEARCH & MANAGEMENT SCIENCE

Frederick S. Hillier, Series Editor Stanford University

Fang, S.-C., Rajasekera, J.R. & Tsao, H.-S.J. / *ENTROPY OPTIMIZATION AND MATHEMATICAL PROGRAMMING*
Yu, G. / *OPERATIONS RESEARCH IN THE AIRLINE INDUSTRY*
Ho, T.-H. & Tang, C. S. / *PRODUCT VARIETY MANAGEMENT*
El-Taha, M. & Stidham , S. / *SAMPLE-PATH ANALYSIS OF QUEUEING SYSTEMS*
Miettinen, K. M. / *NONLINEAR MULTIOBJECTIVE OPTIMIZATION*
Chao, H. & Huntington, H. G. / *DESIGNING COMPETITIVE ELECTRICITY MARKETS*
Weglarz, J. / *PROJECT SCHEDULING: Recent Models, Algorithms & Applications*
Sahin, I. & Polatoglu, H. / *QUALITY, WARRANTY AND PREVENTIVE MAINTENANCE*
Tavares, L. V. / *ADVANCED MODELS FOR PROJECT MANAGEMENT*
Tayur, S., Ganeshan, R. & Magazine, M. / *QUANTITATIVE MODELING FOR SUPPLY CHAIN MANAGEMENT*
Weyant, J./ *ENERGY AND ENVIRONMENTAL POLICY MODELING*
Shanthikumar, J.G. & Sumita, U./*APPLIED PROBABILITY AND STOCHASTIC PROCESSES*
Liu, B. & Esogbue, A.O. / *DECISION CRITERIA AND OPTIMAL INVENTORY PROCESSES*
Gal, T., Stewart, T.J., Hanne, T./ *MULTICRITERIA DECISION MAKING: Advances in MCDM Models, Algorithms, Theory, and Applications*
Fox, B. L./ *STRATEGIES FOR QUASI-MONTE CARLO*
Hall, R.W. / *HANDBOOK OF TRANSPORTATION SCIENCE*
Grassman, W.K./ *COMPUTATIONAL PROBABILITY*
Pomerol, J-C. & Barba-Romero, S. / *MULTICRITERION DECISION IN MANAGEMENT*
Axsäter, S. / *INVENTORY CONTROL*
Wolkowicz, H., Saigal, R., Vandenberghe, L./ *HANDBOOK OF SEMI-DEFINITE PROGRAMMING: Theory, Algorithms, and Applications*
Hobbs, B. F. & Meier, P. / *ENERGY DECISIONS AND THE ENVIRONMENT: A Guide to the Use of Multicriteria Methods*
Dar-El, E./ *HUMAN LEARNING: From Learning Curves to Learning Organizations*
Armstrong, J. S./ *PRINCIPLES OF FORECASTING: A Handbook for Researchers and Practitioners*
Balsamo, S., Personé, V., Onvural, R./ *ANALYSIS OF QUEUEING NETWORKS WITH BLOCKING*
Bouyssou, D. et al/ *EVALUATION AND DECISION MODELS: A Critical Perspective*
Hanne, T./ *INTELLIGENT STRATEGIES FOR META MULTIPLE CRITERIA DECISION MAKING*
Saaty, T. & Vargas, L./ *MODELS, METHODS, CONCEPTS & APPLICATIONS OF THE ANALYTIC HIERARCHY PROCESS*
Chatterjee, K. & Samuelson, W./ *GAME THEORY AND BUSINESS APPLICATIONS*
Hobbs, B. et al/ *THE NEXT GENERATION OF ELECTRIC POWER UNIT COMMITMENT MODELS*
Vanderbei, R.J./ *LINEAR PROGRAMMING: Foundations and Extensions, 2nd Ed.*
Kimms, A./ *MATHEMATICAL PROGRAMMING AND FINANCIAL OBJECTIVES FOR SCHEDULING PROJECTS*
Baptiste, P., Le Pape, C. & Nuijten, W./ *CONSTRAINT-BASED SCHEDULING*
Feinberg, E. & Shwartz, A./ *HANDBOOK OF MARKOV DECISION PROCESSES: Methods and Applications*
Ramík, J. & Vlach, M. / *GENERALIZED CONCAVITY IN FUZZY OPTIMIZATION AND DECISION ANALYSIS*
Song, J. & Yao, D. / *SUPPLY CHAIN STRUCTURES: Coordination, Information and Optimization*
Kozan, E. & Ohuchi, A./ *OPERATIONS RESEARCH/ MANAGEMENT SCIENCE AT WORK*
Bouyssou et al/ *AIDING DECISIONS WITH MULTIPLE CRITERIA: Essays in Honor of Bernard Roy*
Cox, Louis Anthony, Jr./ *RISK ANALYSIS: Foundations, Models and Methods*
Dror, M., L'Ecuyer, P. & Szidarovszky, F. / *MODELING UNCERTAINTY: An Examination of Stochastic Theory, Methods, and Applications*
Dokuchaev, N./ *DYNAMIC PORTFOLIO STRATEGIES: Quantitative Methods and Empirical Rules for Incomplete Information*
Sarker, R., Mohammadian, M. & Yao, X./ *EVOLUTIONARY OPTIMIZATION*
Demeulemeester, R. & Herroelen, W./ *PROJECT SCHEDULING: A Research Handbook*

TRAFFIC THEORY

by

Denos C. Gazis

KLUWER ACADEMIC PUBLISHERS
Boston / Dordrecht / London

Distributors for North, Central and South America:
Kluwer Academic Publishers
101 Philip Drive
Assinippi Park
Norwell, Massachusetts 02061 USA
Telephone (781) 871-6600
Fax (781) 871-9045
E-Mail: kluwer@wkap.com

Distributors for all other countries:
Kluwer Academic Publishers Group
Post Office Box 322
3300 AH Dordrecht, THE NETHERLANDS
Telephone 31 786 576 000
Fax 31 786 546 474
E-mail: services@wkap.nl

Electronic Services <http://www.wkap.nl>

Library of Congress Cataloging-in-Publication Data

Gazis, Denos C., 1930-
Traffic theory / by Denos C. Gazis.
p. cm. -- (International series in operations research & management science ; 50)
Includes bibliographical references and index.
ISBN 1-4020-7095-0
1. Traffic flow. I. Title. II. Series.

HE336.T7 G39 2002
388.3'14'01--dc21

2002067819

Printed on acid-free paper.

Printed in the United States of America.

To my wife, Jean

CONTENTS

Preface

"Everything should be made as simple as possible—but not simpler"
Albert Einstein

Traffic Theory, like all other sciences, aims at understanding and improving a physical phenomenon. The phenomenon addressed by Traffic Theory is, of course, automobile traffic, and the problems associated with it such as traffic congestion. But what causes congestion?

Some time in the 1970s, Doxiades coined the term "oikomenopolis" (and "oikistics") to describe the world as man's living space. In Doxiades' terms, persons are associated with a living space around them, which describes the range that they can cover through personal presence. In the days of old, when the movement of people was limited to walking, an individual oikomenopolis did not intersect many others. The automobile changed all that. The term "range of good" was also coined to describe the maximal distance a person can and is willing to go in order to do something useful or buy something. Traffic congestion is caused by the intersection of a multitude of such "ranges of good" of many people exercising their range utilisation at the same time. Urban structures containing desirable structures contribute to this intersection of "ranges of good".

In a biblical mood, I opened a 1970 paper entitled "Traffic Control -- From Hand Signals to Computers" with the sentence: "In the beginning there was the Ford". The Ford was indeed the beginning of man's modern age of transportation, and the enlargement of individual oikomenopolis, and this beginning took place shortly after the beginning of the 20th century. For a few decades after that, people exploited the capabilities of the new medium by trial and error, (and I am suppressing here the urge to add "mostly error"). The Ford, and its follow up companion cars, increased the mobility of people and gave birth to urban sprawl and its concomitant traffic congestion. Traffic signals, borrowed from railroad train management technology, were brought into the scene to improve safety and combat congestion, with mixed results due to inadequate understanding of the inner workings of automobile traffic. And then Traffic Theory was born.

Around the middle of the 20th century, some notable scientists laid the foundations of understanding of automobile traffic. You can find the names of some of these scientists in an article I wrote for the 50th anniversary issue of Operations Research, entitled "The Origins of Traffic Theory". Much work followed after those pioneering contributions, some of it productive and some readily classified as "solutions in search of a problem". Traffic Theory followed the tradition of other physical sciences, with many scientists entering the field in order to apply what they knew to model traffic. Some of them did not make the necessary effort to justify their modeling effort in terms of realistic models of human behaviour, often ignoring the admonition of Albert Einstein shown at the beginning of this preamble. Comments on such misguided effort will be made later in this book. Suffice it to say here that mere good fit with data is not sufficient to justify a theory, which must also be based of a rational consideration of physical properties of the things that the theory purports to describe. This book concentrates on theoretical work tested over the years, which is based on such foundations of proper pursuit of science.

Traffic theory, after modelling of traffic movement, moved into modelling of interactions of traffic movements aimed at improving them through appropriate control of traffic movement through devices such as traffic signals or other means of regulating the rate of flow of traffic past a control point. Thus, traffic control theory for congested or uncongested systems was developed, once more with mixed success in producing the desired improvement of traffic flow.

If we were to evaluate the overall effect of progress in traffic theory on traffic management, we would find "mixed success" as the universal measure of this effect. This is so in spite of considerable effort made in recent years to utilise advanced computer technology in order to improve the management of transportation systems, under the label of "Intelligent Transportation Systems (ITS)". A short explanation of why success is mixed may be the following:

a) Some good traffic theoretical models are inadequately used.
b) Some other good theoretical models are missing, and inadequate attention is paid toward their development.
c) Some of the suggested solutions to the traffic problem are aptly described by the edict of Harold Mencken : "*For every problem there is a solution that is simple, neat, and wrong*".

In this book, I present key models of traffic flow and associated traffic phenomena such as conflicts in traffic, traffic generation and assignment, and traffic control. The book builds on my earlier work, (**Traffic Science,* edited by Denos C. Gazis, Wiley Interscience, New York, 1974), and the work of other transportation scholars. I have focused on identifying the appropriate use of the various models developed over the years for the improvement of traffic systems. I discuss how proper use of some of these models may accelerate the successful deployment of ITS, and also enumerate opportunities for development of additional models needed for continued improvement of ITS. The book comprises four chapters, namely:

1. Traffic Flow Theories: A description of the movement of traffic and the interaction of its component vehicles with each other.
2. Delay Problems at Isolated Intersections: A description of queuing and delays at isolated intersections.
3. Traffic Control: A discussion of algorithms for optimum control of single intersections and systems of intersections, including networks.
4. Traffic Assignment: A discussion of traffic generation, distribution and assignment in a transportation network.

I would like to express my gratitude to Walter Helly, Donald McNeil and George Weiss, and the late Leslie Edie, whose contributions to the aforementioned book "Traffic Science" provided the basis for the corresponding chapters of this book.

The book is intended as a textbook for a college Transportation Science curriculum, and as a reference book for researchers in Transportation Science. I have concentrated in presenting the fundamental concepts and methods in the various areas of traffic theory, but not necessarily covering all the contributions, which do not alter substantially these concepts and/or methods. Some of these contributions are listed in the References section of each chapter as “additional references”.

Denos Gazis

Chapter 1

Traffic Flow Theories

Traffic flow theories may be viewed as the foundation of Traffic Science. They are intended to provide understanding of phenomena related to the movement of individual vehicles along a highway as they interact with neighbouring vehicles. It is the consequences of such interactions that determine the fundamental characteristics of highways, such as their capacity and their ability to sustain various levels of traffic flow.

Traffic flow theories were among the earliest contributions to Traffic Science, and they have included both a *Macroscopic,* and a *Microscopic* treatment. The macroscopic treatment views traffic as a continuum akin to a fluid moving along a duct which is the highway. The microscopic treatment considers the movement of individual vehicles as they interact with each other.

1.1 MACROSCOPIC TRAFFIC FLOW THEORY – KINEMATIC WAVES IN TRAFFIC

Lighthill and Whitham (1955) laid the foundation of macroscopic traffic flow theory in their seminal paper describing traffic as a fluid moving along a duct. They discussed traffic along a reasonably crowded road, with no appreciable gaps between individual vehicles. In this case, traffic may be viewed as a continuum, and its characteristics correspond to the physical characteristics of the imaging fluid. At this point, some definitions are in order:

q = the "flow rate" of traffic along a highway lane, in vehicles per hour.
u = the (average) speed of the traffic, in miles per hour.
k = the "density" of traffic, in vehicles per mile.

The above quantities satisfy the relationship

$$q = u\,k\,. \tag{1}$$

At this point, we should discuss how these quantities are measured. Basically, traffic is observed in two ways. First, observations may be made at two or more fixed points along a roadway, with measurements taken at the time of occurrence of traffic events at each of these points. Events, in this case, are the arrivals of some parts of vehicles, e.g. the front, at each of the observation points. Second, observations may be made at two or more instants of time. The events observed at each instant are the positions of each vehicle on the roadway. The first way uses a short segment of the roadway, dx, and a large interval of time T; the second uses a short interval of time, dt, and a large segment of the roadway X. This is illustrated in Figure 1, where the trajectories of some 25 vehicles are shown in the space versus time domain, with some of these trajectories traversing the X versus T rectangular domain.

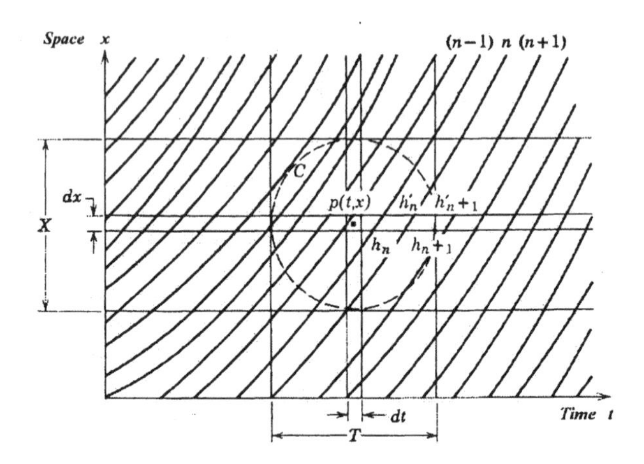

Figure 1. Vehicle space-time trajectories and methods of measurement.

If $N = \Sigma\, n_i$ is the total number of vehicles traversing dx in time T, with n_i at speed u_i, we can compute the values of the three traffic variables by looking at the vehicles as distributed either in time or in space. First, considering the vehicles as distributed in time, and remembering that in this case the average speed of these vehicles must be their harmonic average, we obtain

$$q = \frac{N}{T}$$
$$u = \frac{N}{\sum n_i / u_i}$$
$$k = \frac{q}{u} = \frac{\sum n_i / u_i}{T}$$

By substituting $u_{ii} = dx / t_i$, where t_i is the travel time across dx, and by multiplying the expression for q by dx in numerator and denominator, we obtain

$$q = \frac{Ndx}{Tdx}$$
$$u = \frac{Ndx}{\sum n_i t_i} \qquad (2)$$
$$k = \frac{\sum n_i / t_i}{Tdx}$$

Now, assuming a space distribution of vehicles observed on a stretch of length X over time dt, we let $N = \sum n_i$ be the total number of vehicles observed and remember that in this case the mean speed is the space mean. We thus obtain the following values of the three variables:

$$k = \frac{N}{X}$$
$$u = \frac{\sum n_i u_i}{N}$$
$$q = uk = \frac{\sum n_i u_i}{X}$$

Substituting x_{ii} / dt for u_i, where x_i is the distance traveled by vehicles at speed u_i in the time interval dt, we obtain

$$k = \frac{Ndt}{Xdt}$$
$$u = \frac{\sum n_i x_i}{Ndt} \qquad (3)$$
$$q = \frac{\sum n_i x_i}{Xdt}$$

A comparison of Eqs. 2 and 3 shows that they are equivalent, and lead to the following general definitions for the three traffic variables:

The mean flow q of vehicles travelling over a section of a roadway of length X during a time interval T is the total distance travelled on the roadway by all vehicles which were on this section during any part of time T divided by XT, the area of the space-time domain observed.

The mean concentration k of vehicles travelling over a roadway section of length X during a time T is the total time spent by all the vehicles on X during time T divided by XT.

The mean speed u of vehicles travelling over a roadway section of length X during a time interval T is the total distance travelled on the section by all vehicles that were on it for any part of time T divided by the total time spent by all vehicles on the section during time T.

These definitions may be further generalised by using an area of any shape of the space-time domain, such as a circular or elliptical, rather than the rectangular one shown in Figure 1. In any case, these definitions make sense if the trajectories of the vehicles do not deviate very much from a steady state situation corresponding to a bunch of parallel trajectories.

In their discussion of traffic phenomena, Lighthill and Whitham discussed essentially transitions between steady state situations. They further assumed that q and k were related in a fashion described by what came to be known as the "Fundamental Diagram", shown in Figure 2. The rationale for the existence of a Fundamental Diagram is simple. Clearly, at zero density ($k=0$), there is no traffic and consequently $q=0$. Traffic flow is also zero at some "jam density", k_j, when cars are stopped bumper-to-bumper. Assuming a continuous relationship between q and k, by Rolle's theorem

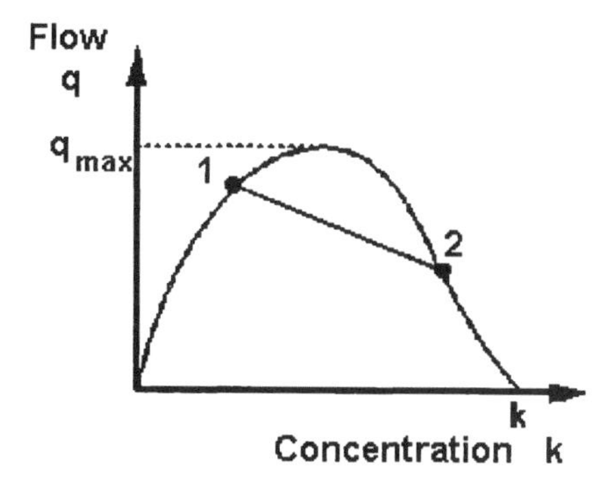

Figure 2. The Fundamental Diagram

there must be at least one maximum point between *0* and k_j . The assumption of one maximum point produces the Fundamental Diagram of Figure 2, which describes the following behaviour of traffic. As the density increases from zero, the flow also increases, starting with an almost linear rate of increase and a lower rate as the density increases and the average speed of the traffic, shown in Figure 3, decreases. The flow reaches a maximum at some optimum speed and density, and then starts decreasing with increasing density, reaching zero at jam density, when the speed is also zero.

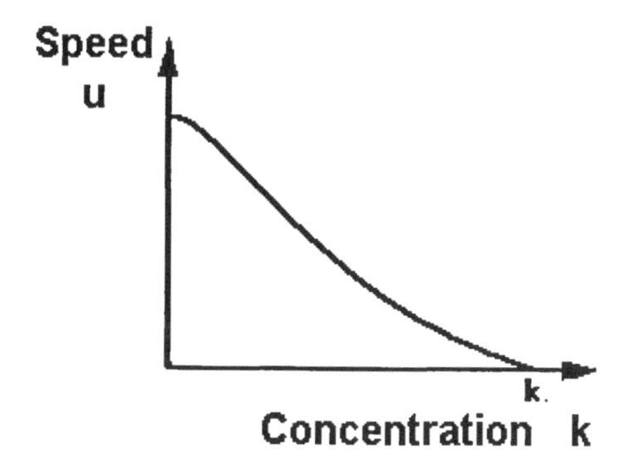

Figure 3 . Speed versus concentration

Finally, Lighthill and Whitham stipulated a principle of conservation of vehicles, expressed by the relationship

$$\frac{\partial k}{\partial t}+\frac{\partial q}{\partial x}=0 \tag{4}$$

Assuming $q = q\,(k)$, we derive from Eq. 4 the following relationship

$$\frac{\partial k}{\partial t}+V\frac{\partial k}{\partial x}=0 \tag{5}$$

where $V = \partial q / \partial k$ is the slope of the tangent at the corresponding point on the Fundamental Diagram. The solution of Eq. 5 is

$$k = F\,(x - V\,t) \tag{6}$$

where F is an arbitrary function. Equation 6 implies that inhomogeneities such as changes in concentration of cars propagate along a stream of traffic at constant speed $V = \partial q / \partial k$, which is positive or negative with respect to a stationary observer, depending on whether the concentration is below or above the optimum concentration corresponding to maximum q (Figure 2). For example, the speed V will be positive for point 1 in Figure 2, and negative for point 2.

1.1.1 Shock waves.

An important derivative of the above theory is the discussion of the propagation of disturbances when traffic changes from one steady state to another. Let us assume, for example, that traffic conditions change by a shift from concentration k_1 to concentration k_2, (Figure 4). The transition

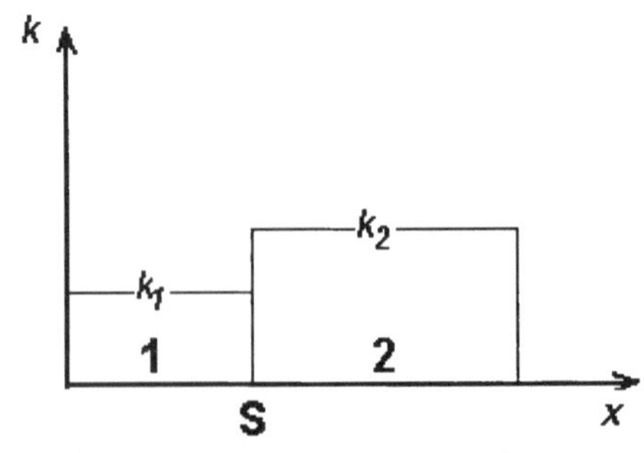

Figure 4. Transition from one concentration to another.

takes place at point S, which in general moves along the roadway, and we are about to find out how it does so. The corresponding points on the flow versus concentration curve are points 1 and 2 in Figure 2. We can visualise the movement of individual vehicles corresponding to these two points as shown in Figure 5, where the trajectory of each vehicle transitions from a straight line with slope u_1 to one with slope u_2. The movement of point S of Figure 4 is represented by the line separating the domains 1 and 2 in Figure 5. The speed of propagation of the "*shock wave*" associated with

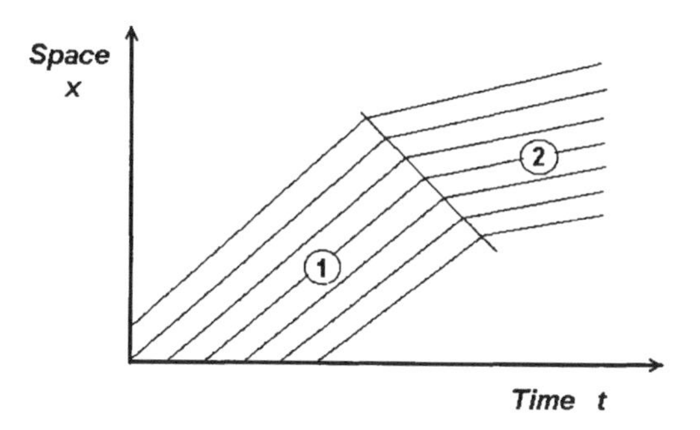

Figure 5. Change of speed of a moving platoon

the change from state 1 to state 2 is the slope of this line. If it is positive, the shock propagates forward with respect to a stationary observer. If negative, the shock propagates backwards. If we denote this speed of propagation of the shock wave as u_w, we can compute its value as follows.

Let us use the notation

u_1 = space mean speed of vehicles in region 1
u_2 = space mean speed of vehicles in region 2
$U_{r1} = (u_1 - u_w)$ = speed of vehicles in region 1 relative to the moving line S (Fig. 4).
$U_{r2} = (u_2 - u_w)$ = speed of vehicles in region 2 relative to the moving line S.

The number N of vehicles crossing the dividing line S is

$$N = U_{r1}\, k_1\, t = U_{r2}\, k_2\, t$$

which leads to the relationship

$$(u_1 - u_w) k_1 = (u_2 - u_w) k_2 \tag{7}$$

Equation 7 is a restatement of the principle of conservation of vehicles, and may be rewritten in the form

$$u_2 k_2 - u_1 k_1 = u_w (k_2 - k_1) \tag{8}$$

and using Equation 1, ($q = u k$),

$$u_w = \frac{(q_2 - q_1)}{(k_2 - k_1)} \tag{9}$$

It may be observed that Equation 9 can be obtained directly on geometric grounds from Figure 5, which obviously satisfies the principle of conservation of vehicles. It is the fundamental equation for the speed of propagation of "kinematic" waves in traffic according to Lighthill and Whitham.

At this point, a few observations are in order regarding the underlying assumptions in the derivation of Equation 9, and corresponding limitations in its applicability. First of all, it is clear that a vehicle does not change its speed abruptly from u_1 to u_2 as shown in Figure 5, but rather there is a gradual transition from one speed to another. As long as this transition is very similar among different vehicles, the derivation of Equation 9 is reasonably valid in describing something akin to a shock generated by the transition from one steady-state flow situation to another. However, one should be careful not to use Equation 9 in describing a succession of rapid transitions from one state of flow to another, when the vehicles do not have a chance to relax to a steady-state flow before entering another transition. What complicates matters in real life is the fact that the Lighthill-Whitham theory inherently describes traffic in an infinitely long, flat, and straight roadway. The existence of geometric variations in the roadway influences the behaviour of traffic in all respects, including the propagation of kinematic waves. However, Equation 9 does provide a reasonably good description of the shock wave generated when traffic meets a bottleneck, as indicated in Figures 2 and 5, where the flow q_2 downstream cannot accommodate the flow q_1 upstream. Moreover, it may be observed that the derivation of Equation 9 does not require the existence of a unique flow versus concentration relationship. It can be derived on geometric grounds from Figure 5, and only presupposes the existence of two possible steady-state flow conditions, (q_1 , k_1) and (q_2 , k_2).

1.1.2 Flow versus concentration relationships

The preceding discussion applies to any flow versus concentration relationship, and the question naturally arises which such relationship is the appropriate one. Several such relationships have been proposed over the years, and some have been tested against experimental evidence.

Richards (1956) proposed a hydrodynamic model of traffic essentially identical to that of Lighthill and Whitham, independently of them. He also proceeded to explore the properties of his model assuming a specific relationship between speed and density, namely,

$$u = u_f \left(1 - \frac{k}{k_j} \right) \tag{10}$$

which leads to a parabola for the q versus k curve,

$$q = u_f \left(k - \frac{k^2}{k_j} \right) \tag{11}$$

Richards used this flow-concentration relationship explicitly in all his discussions of traffic phenomena, thus limiting the applicability of his results.

The first model for the q-k curve derived entirely on the basis of a hydrodynamic model was that proposed by Greenberg (1959), who assumed that traffic could be approximated by a one-dimensional fluid satisfying the continuity equation. He assumed an equation of motion for the fluid, which

$$\frac{du}{dt} = \frac{-a^2}{k} \frac{\partial k}{\partial x} \tag{12}$$

states that the acceleration of the traffic stream is inversely proportional to the concentration k and directly proportional to the concentration gradient $\partial k / \partial x$, with a constant of proportionality of $-a^2$. If the speed is given as a function of location and time, i.e. $u = u(x, t)$, the equation of motion becomes

$$\frac{\partial u}{\partial t} + u\frac{\partial u}{\partial x} + \left(a^2 / k\right)\frac{\partial k}{\partial x} = 0 \qquad (13)$$

Now, the continuity equation, (Equation 4), together with the relationship $q = uk$, yield

$$\frac{\partial k}{\partial t} + u\frac{\partial k}{\partial x} + k\frac{\partial u}{\partial x} = 0 \qquad (14)$$

If it is now assumed that the speed of the traffic stream is only a function of the concentration, i.e. $u = u(k)$, then

$$\begin{aligned} \frac{\partial u}{\partial x} &= \frac{du}{dk}\frac{\partial k}{\partial x} \\ \frac{\partial u}{\partial t} &= \frac{du}{dk}\frac{\partial k}{\partial t} \end{aligned} \qquad (15)$$

Substituting Equations 15 into Equations 13 and 14, and solving for du/dk, we obtain

$$\frac{du}{dk} = -\frac{a}{k}$$

Integrating, and using the end condition $u = 0$ when $k = k_j$, we obtain

$$
\begin{aligned}
u &= a\ln\left(\frac{k_j}{k}\right) \\
q &= ku = ak\ln\left(\frac{k_j}{k}\right)
\end{aligned}
\tag{16}
$$

The maximum q is given by Equation 16 at $k_j / k = e$, when the corresponding speed at maximum flow, u_m, is equal to a, yielding finally

$$
\begin{aligned}
a &= u_m \\
q &= u_m k\ln\left(\frac{k_j}{k}\right)
\end{aligned}
\tag{17}
$$

A plot of q versus k according to Equation 17 is shown in Figure 6, together with field measurements taken at the Lincoln Tunnel of New York City, as discussed by Edie (1974). There is absence of data for very low and very high densities, and a scatter of data, particularly past the point of maximum flow, but overall the fit is reasonably satisfactory.

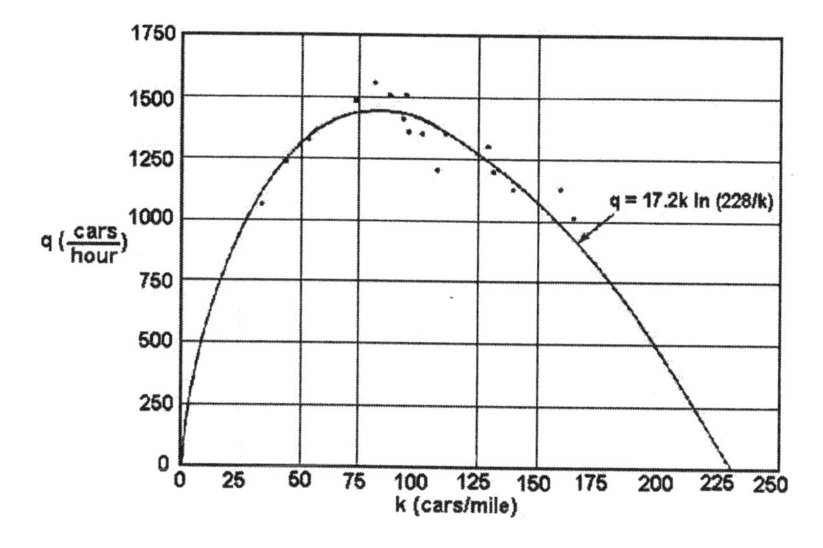

Figure 6. Greenberg's model of flow versus concentration, together with field data from the Lincoln Tunnel of NYC

There are many possible causes for the scatter of data at densities corresponding to relatively dense traffic. Some of these causes are discussed in what follows.

1.1.3 Additional considerations of the macroscopic traffic flow theory

As mentioned above, the representation of the properties of a roadway by a single flow versus concentration relationship applies, at best, on an infinitely long, flat, straight roadway. The existence of entrance and exit ramps, curves, and slopes introduces variations, which may be handled by assuming concatenation of roadway sections with different flow versus concentration relationships. This will generate the appropriate transitions from one q,k state to another, including the emergence of bottlenecks.

In addition to geometry, some special considerations of driver behaviour may also dictate a departure from a single q versus k relationship. Newell (1965) was the first one to consider the effect of different driver behaviour during acceleration and deceleration. It turns out that all drivers generally allow greater gaps between cars during acceleration than they do during deceleration. This is not surprising, because drivers may drive close to the minimum distance for safe driving during deceleration, but they may relax and let gaps increase for a while during acceleration. This introduces, inherently, lower values for flow during acceleration than during deceleration, for the same value of density. Newell assumed that the properties of the roadway during acceleration were described by a different q versus k relationship, and defined this effect as an "acceleration bottleneck". The net effect of this asymmetry in acceleration and deceleration behaviour is that if a car, and its followers, undergo a brief deceleration manoeuvre and then go back to their original speed, the speed recovery will be associated with a decrease in density and flow. This will propagate indefinitely upstream, causing effects similar to those of passing through a bottleneck of decreased capacity, hence the name of "acceleration bottleneck". The phenomenon has been described as a "hysteresis effect", altering the transition from one q,k state to another, as shown in Figure 7. In this figure, it is assumed that a car and its followers undergo a deceleration manoeuvre from point 1 to point 2. The hysteresis effect is described by the dashed curve, emanating from point 2. As the cars try to attain their original speed, they end up not back to point 1 but to point 3, generating a shock wave that affects all the cars behind them.

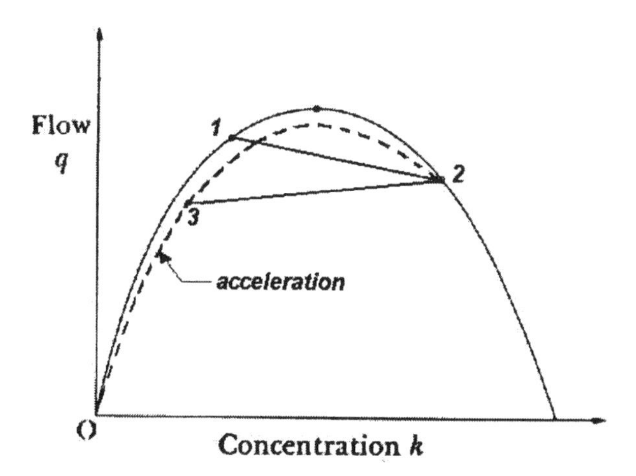

Figure 7. Shock wave caused by asymmetry between deceleration and acceleration.

There is considerable experimental evidence supporting Newell's model of the asymmetry between acceleration and deceleration. Forbes et al (1958) have reported slower response of drivers to acceleration than to deceleration. Herman and Potts (1961) reported measurements indicating differences in behaviour during acceleration ad deceleration. Daou (1964) found significantly larger spacing between cars accelerating at a given speed than between cars decelerating at the same speed. In addition, there have been observations through aerial photographs showing the existence of accelerating shock waves.

The deceleration may be due to the presence of a curve along the roadway necessitating a temporary decrease in speed, followed by acceleration to the original speed past the curve. This shows how geometry, coupled with driver behaviour, may alter the behaviour of traffic beyond the predictions of the Lighthill-Whitham treatment. An even more serious effect of geometry is that caused by the existence of grades along the roadway. In particular, an uphill portion may cause a large disturbance in traffic, particularly if a heavy vehicle such as a truck must negotiate this uphill portion together with the rest of the traffic because no "slow traffic" lane is provided along the uphill portion of the roadway. If the truck slows down sufficiently, it may block not only its own lane but also the neighbouring lane, since cars trapped behind the truck try to escape into this neighbouring lane, thus impeding the movement of cars along that lane as well. In a two-lane roadway, the net effect is akin to that caused when two lanes merge into one, the only difference being that the merging point moves with the speed of the slow truck. This produces the effect of a "moving bottleneck" described by Gazis and Herman (1992). The moving bottleneck may be transformed into a "phantom bottleneck" if the truck reaches a level or downhill portion and resumes its normal speed. In that case, drivers trapped behind the moving bottleneck see no bottleneck when they move past the congested portion caused by the moving bottleneck and see no apparent cause for this congestion.

A brief discussion of the study of the moving and phantom bottleneck will be given here, to illustrate the use of kinematic models for the study of complex traffic phenomena.

As mentioned already, it is assumed that the moving bottleneck is caused by a vehicle moving at a speed v_1 substantially slower than the average speed of the traffic stream, v_0. We assume that it moves along the right lane of a two-lane stream of traffic. Because of its reduced speed, cars queue up behind it, driving at the same slow speed and waiting for an opportunity to pass. By manoeuvring for passing, they slow down the neighbouring lane as well, roughly to the same slow speed, resulting in a two-lane queue, which is discharged at the front of the left (unblocked) lane. The situation resembles the merging of two lanes into one, only the point of convergence moves at the speed of the slow vehicle, the moving bottleneck.

When a vehicle finds itself in the unblocked lane abreast of the slow vehicle, it escapes at a speed v_2 which is generally different from both v_1 and v_0 Let us assume that a vehicle escaped at time $t = 0$ (Figure 8). After a time Δt, it will travel a distance $v_2 \Delta t$, while the slow vehicle travels a distance $v_1 \Delta t$. If the spacing between escaping vehicles is s_2, then the flow past the slow vehicle is

$$Q_e = \frac{1}{\Delta t} \frac{(v_2 \Delta t - v_1 \Delta t)}{s_2} = \frac{(v_2 - v_1)}{s_2} \tag{18}$$

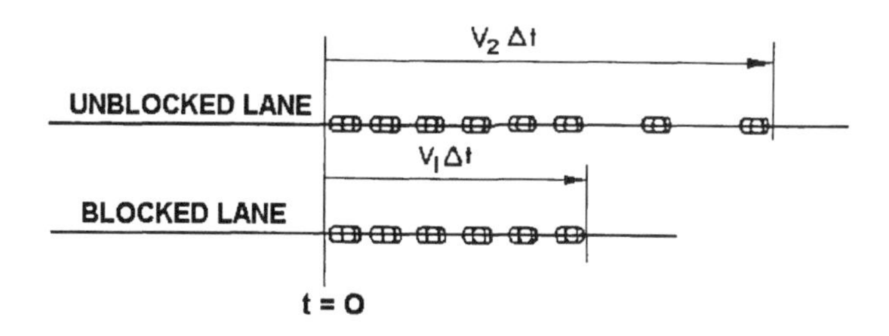

Figure 8. Behaviour of vehicles in the vicinity of a moving bottleneck.

According to our assumptions, the slow vehicle leads a pack alongside and behind it moving at speed v_1 . The front of this pack will travel a distance $v_1\Delta t$ during the time interval Δt. It follows that the slow vehicle leads a two lane pack which supplies an additional flow rate, Q_a , equal to twice that corresponding to speed v_1 and concentration k_1 , namely,

$$Q_a = 2q_1 = 2\frac{v_1}{s_1} = 2k_1 v_1 \tag{19}$$

The total flow along the two-lane highway is

$$Q = Q_e + Q_a = 2k_1 v_1 + (v_2 - v_1)k_2 \tag{20}$$

Equation 20 has all the desired properties in the limits $v_1 = 0$ and $v_1 = v_2$, and is based on reasonable assumptions regarding the effect of the slow vehicle on other vehicles during periods of moderately heavy traffic. Next, we can estimate the speed of the trapped vehicles behind ad alongside the slow vehicle. This is not equal to the speed of the slow vehicle, because a portion of the two queues escapes. Assuming that the percentage of escaping vehicles is equally divided among the two lanes, we use the definition $v = q/k$ and the value of flow given by Equation 20, with concentration k_1 in each of the two lanes, obtaining the speed of the trapped vehicles

$$v_3 = \frac{Q}{2k_1} = v_1 + (v_2 - v_1)\frac{k_2}{2k_1} \tag{21}$$

We can now determine whether or not the queue size increases or decreases by using the Lighthill-Whitham estimation of the velocity of a shock wave. The queue size remains constant if the speed of the front of the queue with respect to a stationary observer is equal to that of the rear of the queue. The speed of the front of the queue is obviously equal to that of the slow vehicle, namely, v_1. The speed of the end of the queue is

$$v_s = \frac{\Delta q}{\Delta k} = \frac{\frac{1}{2}Q - q_0}{k_1 - k_0} \tag{22}$$

Setting $v_1 = v_s$, and using the value of Q given by Equation 20, we obtain

$$v_1 = \frac{\frac{1}{2}q_2 - q_0}{\frac{1}{2}k_2 - k_0} \tag{23}$$

The value of v_1 given by Equation 23 is a critical value of the speed of the slow vehicle. Above this speed, the slow vehicle has virtually no effect on the overall quality of traffic, because the queue length that might get somehow started will keep decreasing until it is dissipated. Below that critical speed, the queue will tend to increase indefinitely. It may also be pointed out that the preceding treatment does not provide a means for determining the stationary size of a queue, since any size will remain constant if v_1 satisfies Equation 23. It is very possible that, if a moving bottleneck is suddenly introduced into a stream with a speed v_1 satisfying Equation 23, a non-zero queue size will be developed, which depends on the exact movement of cars around the slow vehicle, after which the queue size will remain more or less constant.

1.2 THE BOLTZMANN-LIKE MODEL OF TRAFFIC

Around 1959, Ilya Prigogine, who was later to receive a Nobel Prize for his work on Statistical Mechanics, proposed a treatment of traffic on a multilane highway analogous to the Boltzmann approach for deriving the properties of gases from consideration of the movement of individual atoms of gas. Prigogine (1961), and Anderson et al (1962) developed a theory of traffic flow based on a description of traffic in terms of a probability density for the speed, v, of an individual car. This density, $f(x, v, t)$, may vary as a function of time, t, and a co-ordinate x along the highway. The basic equation for the distribution is assumed to be

$$\frac{\partial f}{\partial t} + v\frac{\partial f}{\partial x} = \left(\frac{\partial f}{\partial t}\right)_{relaxation} + \left(\frac{\partial f}{\partial t}\right)_{interaction} \tag{24}$$

It is assumed that every vehicle enters the highway with a "desired speed", and drives at that speed as long as it is not impeded by a slower car. If it does get impeded, it slows down until it gets a chance to pass, after which it relaxes back to its desired speed. The first term of the right-hand side of Eq. 24 is a consequence of the fact that $f(x, v, t)$ differs from some "desired" speed distribution $f^0(v)$, and drivers try to "relax" to their desired speed as soon as they have an opportunity. The second term of the right-hand side corresponds to the slowing down of a fast vehicle caused by a low one. True to his tradition as a leader in statistical mechanics of fluid media, Prigogine frequently referred to this second term as a "collision" term, a rather unsettling choice of words when speaking about traffic.

The form of these two terms was chosen for mathematical convenience and plausibility. In the end, Eq. 24 was assumed to have the specific form

$$\frac{\partial f}{\partial t} + v\frac{\partial f}{\partial x} = \frac{f - f^0}{\tau} + (1-p)k(\bar{v} - v)f \tag{25}$$

where τ is the characteristic relaxation time, p is the probability that a car passes another car, and $\bar{v}$ is the average speed of the stream of traffic. The second term of the right-hand element of Eq. 25 corresponds to the interaction term of Eq. 24, and tends to approach zero when the concentration of cars is light and the probability of passing is close to unity. In that case, the relaxation term is dominant. If. In addition, we assume a highway with a homogeneous distribution of cars, then $\partial f / \partial x = 0$, and the solution of Eq. 25 is

$$f(v,t) = f^0(v) + \left[f(v,0) - f^0(v)\right]e^{-t/\tau} \tag{26}$$

If we seek a solution of Eq. 25 that is independent of time and space, then the left-hand element of this equation is zero. If, in addition, we consider small values of concentration, the solution ofEq. 25 shows an approximately linear increase of flow with concentration, namely,

$$q = \bar{v}^0 k \tag{27}$$

where $\bar{v}^0$ is the average of the desired speed. As k increases, the flow q falls below the straight line of Eq. 27 due to the increasing influence of interactions between cars, and the resulting slowing down of the faster cars by the slower ones. Moving to the range of high concentrations, we find[11] that q is independent of f^0 and depends only on τ and p, according to the relationship

$$q = \frac{1}{\tau(1-p)} \tag{28}$$

Equation 28 describes what may be described as the “collective flow” of traffic, which is characterised by high densities and very little passing.

The overall behaviour of traffic is described by the curve on the Figure 9 below.

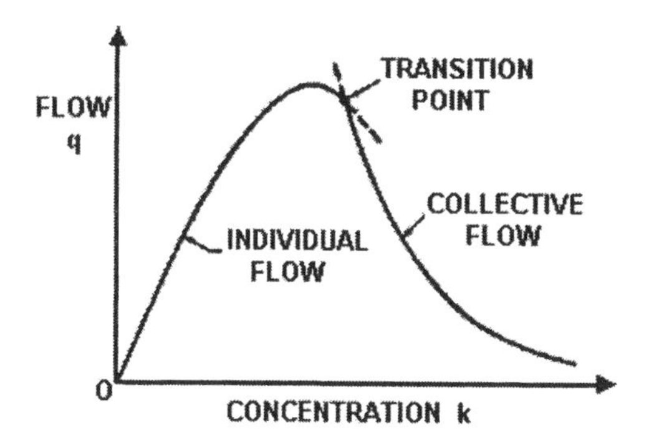

Figure 9. Flow versus concentration according to the Boltzmann-like model of traffic flow

Starting at zero concentration, the flow increases linearly with concentration. The increase slows down as the concentration increases, but the flow remains essentially "individual", meaning that individual vehicles have a chance to drive at their desired speed for a great percentage of the time. At some point, after flow reaches a maximum, the interactions become more and more intensive, until the flow becomes "collective", meaning that individual vehicles are forced to follow other vehicles most of the time. Such behavior clearly matches personal experiences of all drivers, making the Boltzmann-like model quite successful in representing real-life phenomena.

1.2.1 The two-fluid model of traffic

Herman and Prigogine (1979) extended the above discussion, proposing a two-fluid model of traffic flow, which is applicable in the domain of collective flow. According to this model, a portion of the cars are moving, and another portion are "immobilised". At the transition to collective flow, the speed distribution function for cars splits into two parts, one corresponding to the moving vehicles and the other to vehicles stopped as a consequence of congestion, traffic signals, stop signs and barricades, but not in the parked condition.

Extending the ideas in the Boltzmann-like model, they assumed that the average speed of the moving cars, v_R , depended on the fraction of cars that were moving, f_R , according to a power law, namely,

$$v_R = v_m f_R^{\ n} \tag{29}$$

where v_m is the average maximum speed and n is a parameter to be determined from observations. In addition, they stipulated that the fraction of cars stopped, f_s, could be equated to the fraction of the time a car circulating in the street network would be stopped, namely,

$$f_s = T_s / T \tag{30}$$

where T_s is the stopped time per unit distance, T is the trip time per unit distance, and $f_R + f_s = 1$. Equations 29 and 30 lead to a simple transcendental equation relating the stop time to the trip time, both per unit distance, namely,

$$T_s = T - T_m^{1/(n+1)} T^{n/(n+1)} \tag{31}$$

where

$$T_m = \frac{1}{v_m} \tag{32}$$

The two-fluid model has been scrutinised extensively over the years, and found to represent remarkably well the behaviour of traffic during congested periods in various cities. In general, several hours of data were sufficient to determine the two-fluid parameters, namely T_m and n, for a typical urban street network of the central business district of a city. Thus, the two-fluid model provided a means for a quantitative assessment of the quality of traffic in a large number of cities in the United States and abroad. A further investigation and validation of this model was provided by simulation studies of Mahmassani et al (1987). The simulation provided an important supporting tool for further development and extension of the theory, especially the dependence of the proposed relations among network-wide variables on the topological and operational features of a network.

In concluding this section, we may call attention to the following feature of the two-fluid model. In a situation fraught with complexity, involving a multitude of human operators in charge of a variety of vehicles, it is remarkable that the behaviour of traffic can be described by a relatively simple set of equations such as Eqs. 29 to 32. This is because the constraints imposed by traffic congestion limit the number of choices for each individual driver, and force the system into a collective behaviour that is more regimented than one might expect. Obtaining a relatively simple representation of traffic gives, in turn, hope that improvements in traffic management may be implemented by influencing the operational characteristics of a network, once the dependence of performance on these operational characteristics is established through observations and simulation.

1.3 MICROSCOPIC TRAFFIC FLOW THEORY – CAR-FOLLOWING MODELS

The first microscopic models of traffic were proposed, independently, by Reuschel (1950) and by Pipes (1953). They both used models inspired by rules of driving analogous to the well known "California rule", namely, allowing a certain gap between a car and it leader proportional to its speed. Reuschel and Pipes assumed that drivers achieved such goals essentially instantaneously, and they derived the consequences of such behaviour. A more discerning view of driver behaviour was the foundation of the car-following studies carried out by Herman and his collaborators at the General Motors research Labs, starting in 1957. They considered the devices under the control of the driver, which are the brake and the gas pedal. Both control the acceleration of the vehicle, and it was this variable that was the one viewed as controlled by driver action. Second, they took into account the fact that a human being has a time lag in reacting to any inputs. As for the dominant inputs stimulating driver reaction, a combination of rational thinking, mathematical analysis, and experimental observation led to the relative speed between leading and following car as the leading stimulus. The result was the basic car following equation

$$\frac{d^2 x_{n+1}(t+T)}{dt^2} = \lambda \left[\frac{dx_n(t)}{dt} - \frac{dx_{n+1}(t)}{dt} \right] \tag{33}$$

where n denotes the number corresponding to the lead car, $n+1$ that of the following car, T is the reaction time lag, and λ is a coefficient of proportionality of the driver reaction to the stimulus, variously referred to as the sensitivity coefficient or the gain factor. In the first car following studies, this sensitivity coefficient was assumed constant. Herman and his colleagues went on to test their model by conducting car following experiments, and investigating the consequences of this model on the stability of flow of a stream of traffic.

Stability refers to the ability of a system of cars to absorb a perturbation introduced into the stream of traffic. "Local" stability refers to the relative position of a fixed pair of cars. "Asymptotic" stability refers to the attenuation or increase of the amplitude of perturbation as it moves down a line of cars upstream of the perturbation.

1.3.1 Local stability.

Local stability was investigated by solving Eq. 33 by means of Laplace transform techniques (Gazis et al, 1963). Imagine that the lead car undergoes a manoeuvre during $t \leq T$ and then maintains a constant speed for $t > T$. The following vehicle moves according to Eq. 33. If

$$y = \frac{dx_{n+1}}{dt} - \frac{dx_n}{dt} \tag{34}$$

the following equation is obtained for $t > 0$,

$$\frac{dy(t+T)}{dt} = -\lambda y(t) \tag{35}$$

In order that this be a well posed problem, the value of y must be assumed to be given in the interval $0 \leq t \leq T$. To make this point explicit, $y(t)$ may be written as a sum of two functions, one of which is known and the other one unknown, namely,

$$y(t) = y_0(t) + z(t) \tag{36}$$

in which y_0 represents the initial data, given for $0 \le t \le T$, and $z(t)$ represents the unknown part of y, for $t > T$. The functions y_0 and z satisfy the conditions

$$y_0(t) = 0, \quad t > T$$

$$z(t) = 0, \quad t \le T \tag{37}$$

Entering these values of $y_0(t)$ and $z(t)$ in Eq. 35, we obtain

$$\frac{dy(t+T)}{dt} = -\lambda[y_0(t) + z(t)] \text{ , } t > 0 \quad . \tag{38}$$

The Laplace transforms are defined as

$$z^*(s) = \int_0^\infty e^{-st} z(t) dt = \int_T^\infty e^{-st} z(t) dt \tag{39}$$

and

$$y_0^*(s) = \int_0^\infty e^{-st} y_0(t) dt = \int_0^T e^{-st} y_0(t) dt \tag{40}$$

The transform of Eq. 38 is

$$sz^*(s) e^{sT} - y_0(T) = -\lambda[y_0^*(s) + z^*(s)] \tag{41}$$

from which we obtain

$$z^*(s) = -\frac{\lambda y_0^*(s)}{se^{sT}+\lambda} + \frac{y_0(T)}{se^{sT}+\lambda} \tag{42}$$

The denominator of Eq. 42 is now expanded into partial fractions according to the relationship

$$\frac{1}{se^{sT}+\lambda} = \sum_{n=0}^{\infty} \frac{A_n}{(s-s_n)} = \frac{1}{\lambda}\sum_{n=0}^{\infty} \frac{s_n}{1+s_n T}\frac{1}{(s-s_n)} \tag{43}$$

in which s_n is the denumerably infinite set of roots of the equation

$$se^{sT} + \lambda = 0 \tag{44}$$

These roots are assumed to be arranged in ascending order according to the value of their real part, namely,

$$\mathrm{Re}(s_0) > \mathrm{Re}\ (s_1) > \mathrm{Re}(s_2)\ \ldots.$$

The inverse transform of $z^*(s)$ is now found by using the convolution theorem for the first term in Eq. 42, and inverting term by term, and by inverting the second term using the result of Eq. 43. The result is

$$z(t) = \sum_n \frac{b_n s_n}{1+s_n T} e^{s_n t} \tag{45}$$

in which

$$b_n = \int_0^T y_0(t) e^{s_n t} dt - \frac{y_0(T)}{\lambda} \tag{46}$$

It is seen that the long time behaviour of $z(t)$, and hence $y(t)$, is approximately that of the leading term of the series in Eq. 45, namely,

$$y(t) \approx \frac{b_0 s_0}{1 + s_0 T} e^{s_0 t}, \qquad \frac{t}{T} >> 1 \tag{47}$$

in which s_0 is the root of Eq. 44 with the largest real part. The nature of this root, s_0, depends on the value of the product

$$C = \lambda T \tag{48}$$

Thus, it is found that three different types of limiting behaviour of $y(t)$ are possible, depending on the magnitude of $C = \lambda T$, as follows:

If $\lambda T \leq e^{-1}$, then s_0 is real and negative, and therefore the initial fluctuation is damped out monotonically, for sufficiently large values of t.

If $e^{-1} < \lambda T < \pi/2$, the quantity s_0 is complex with a negative real part. Hence, the initial fluctuation is damped out as the product of a negative exponent and a trigonometric function of time. When $\lambda T = \pi/2$, the fluctuation remains constant in amplitude and oscillates sinusoidally in time.

If $\lambda T > \pi/2$, the quantity s_0 has a positive real part. Hence, the initial fluctuation produces oscillations of increasing amplitude, and equilibrium is never reached.

The preceding discussion is illustrated in Figs. 10 and 11 which show the results of numerical computations of the trajectories of the lead and the following car using Eq. 33, for various values of C, and assuming a fluctuation in the acceleration of the lead car in the form of a negative square pulse followed by a positive pulse of the same magnitude. Figure 10 is a graph of the detailed motion of the two cars, showing the effect of the acceleration pattern of the lead car, for different values of C. Remembering that e^{-1}=0.368, we find that for C = 0.5 and 0.8 the spacing is oscillatory and damped, for C=1.57 it is oscillatory and undamped, and for C=1.6 it is oscillatory with increasing amplitude.

1.3.2 Asymptotic stability.

Although a chain of vehicles may be locally stable, it may result in an asymptotically unstable situation, when each vehicle amplifies the signal and passes it on to the cars upstream. This possibility is investigated by considering a Fourier component of the fluctuation of the lead vehicle, the 0th, in a chain. Let

$$\frac{dx_n(t)}{dt} = f_n e^{i\omega t}$$
$$f_0 = 1 \tag{49}$$

Substituting into Eq. 33, we obtain

$$\frac{i\omega}{\lambda} e^{i\omega t} f_{n+1} = f_n - f_{n+1} \tag{50}$$

hence

$$f_n = \left(1 + \frac{i\omega}{\lambda} e^{i\omega t}\right)^{-n} f_0 \tag{51}$$

and

$$x_n(t) = \left[1 + \frac{\omega^2}{\lambda^2} - 2\frac{\omega}{\lambda}\sin\omega t\right]^{\frac{-n}{2}} \exp i\left[\omega t - n\tan^{-1}\left(\frac{\frac{\omega}{\lambda}\cos\omega t}{1 - \frac{\omega}{\lambda}\sin\omega t}\right)\right] \tag{52}$$

The amplitude factor in Eq. 52 decreases with increasing values of n if

$$\frac{\omega^2}{\lambda^2} > 2\frac{\omega}{\lambda}\sin\omega t \tag{53}$$

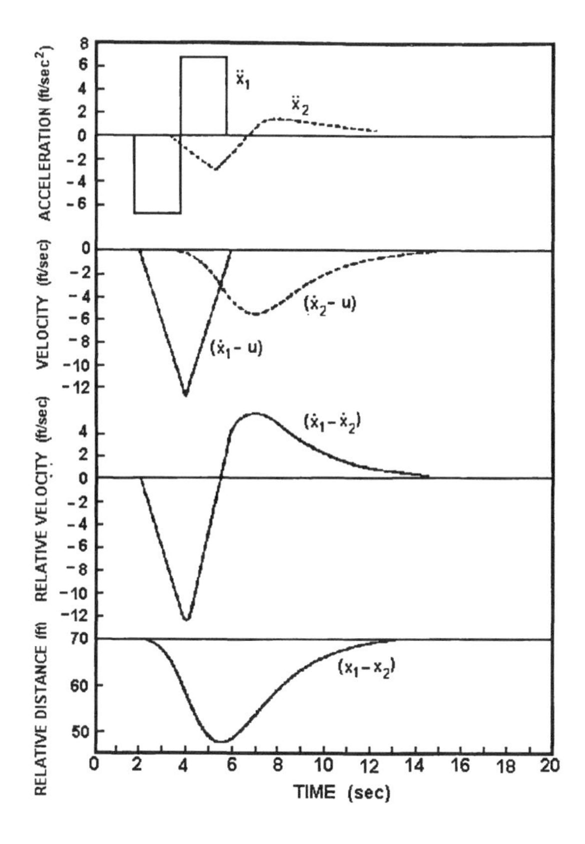

Figure 10. Detailed motion of two cars

It may be seen from Eq. 53 that low frequencies cause the greatest limitations on the sensitivity λ and time lag T. For $\omega \rightarrow 0$, in order to have decreasing amplitude of the perturbation upstream, λ and T must satisfy the relationship

$$\lambda T < \frac{1}{2} \tag{54}$$

Equation 54 gives the condition for asymptotic stability. Numerical examples are given in Figs. 12 and 13.

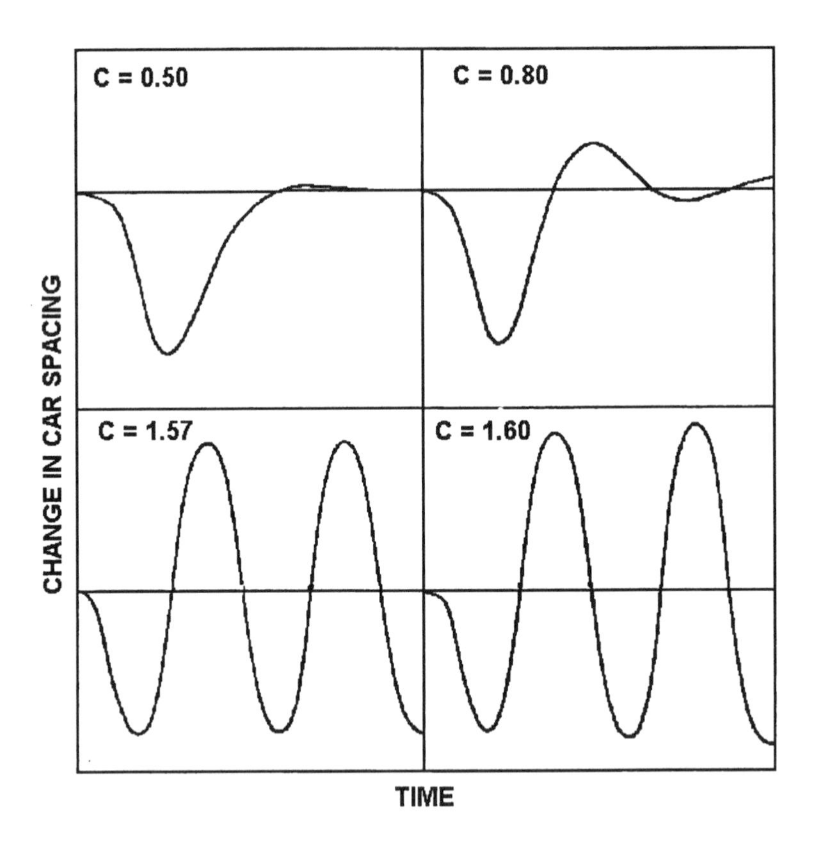

Figure 11. Change in spacing of two cars versus time

Equation 54 gives the condition for asymptotic stability. Numerical examples are given in Figs. 12 and 13.

In Fig. 12, the spacing between successive pairs of cars is shown, after a deceleration-acceleration manoeuvre of the lead car, and for three different values of C.

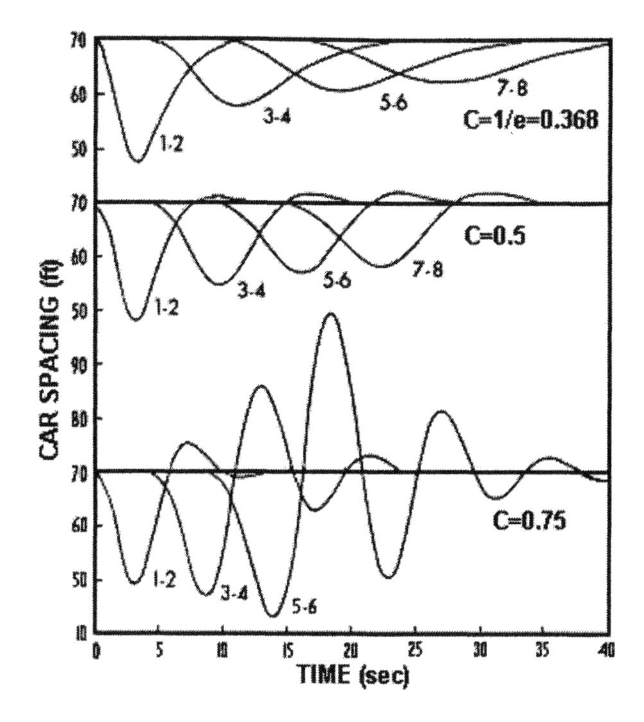

Figure 12. Car spacing between successive pairs of cars after car #1 introduces a perturbation.

In all three cases, we have local stability in the sense that the amplitude of the perturbation in the spacing between cars eventually decreases toward zero with increasing time. However, in the third case, $C = 0.75$, the amplitude of the perturbation increases between successive pairs of cars, causing asymptotic instability. This is shown more dramatically in Fig. 13, where the position of nine cars is shown with respect to a system of co-ordinates that moves with the original speed of the lead car. The original spacing of the cars is 40 ft, and it is assumed that $C = 0.8$ and $T = 2$ sec. It is seen that the spacing between successive pairs of cars suffers a perturbation of increasing amplitude as we move upstream, until there is a collision between the 7th and 8th cars.

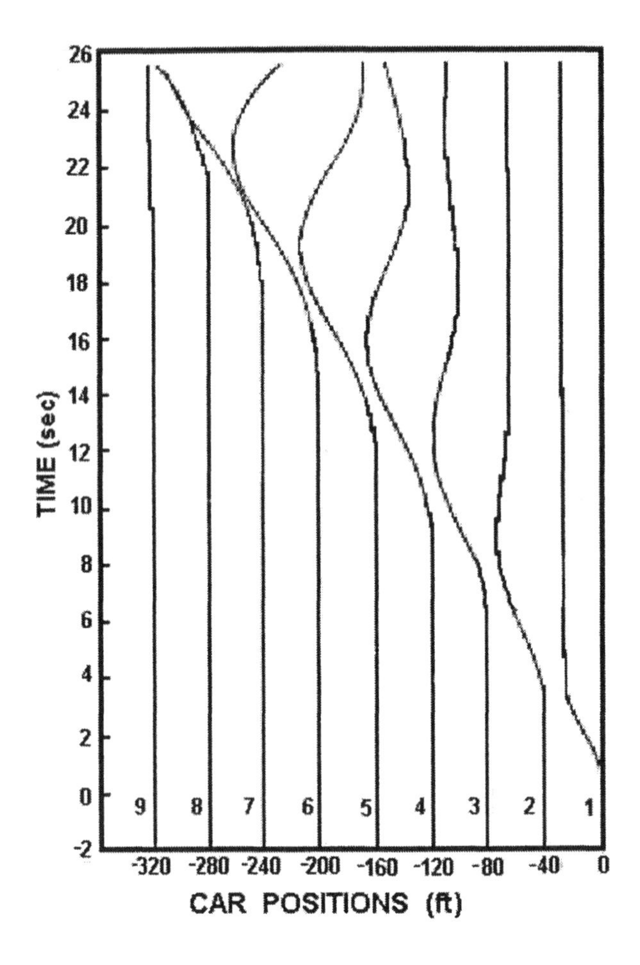

Figure 13. Demonstration of asymptotic instability

It may be mentioned here that the original experiments carried out at General Motors by Herman and his colleagues produced, for eight test drivers, an average value of C equal to 0.56, indicating that people drive on the verge of asymptotic instability. However, there are some additional stabilising factors in real life driving. For example, people are generally able to see more than just the car in front of them, and adjust their driving accordingly. It has been demonstrated analytically (Herman et al, 1959) that taking into account information from the next-nearest neighbour enhances the stability of a stream of traffic. On the other hand, when such next-nearest neighbour information is lacking, for example due to fog or extreme congestion conditions, such stabilising influence may be lacking, which may explain the higher incidence of rear end collisions in such circumstances.

1.3.3 Acceleration noise.

It is reasonable to assume that an average driver would like to maintain a constant speed during a trip, avoiding accelerations and decelerations. However, deviations from a constant speed are generated by a number of factors including the geometry of the roadway, the influence of neighbouring cars, and the driver's own propensity toward aggressive rather than passive driving behaviour. In general, high volumes of traffic tend to push every driver toward car-following patterns of driving, generating greater fluctuations from constant speed than those present at low traffic volumes.

A measure of the velocity fluctuations during a trip is given by the standard deviation σ of the acceleration about a mean acceleration, and has been defined as the "acceleration noise". Assuming the mean acceleration during a trip to be zero, the acceleration noise is defined as

$$\sigma = \left\{(1/T)\int_0^T [a(t)]^2 dt\right\}^{1/2} \quad , \tag{55}$$

where $a(t)$ is the acceleration (positive or negative) at time t, and T is the total time of the trip. A difference approximation to Eq. 55, in which acceleration is assumed to be sampled at time intervals (Δt), is given by

$$\sigma = \left\{(1/T)\sum [a(t)]^2 \Delta t\right\}^{1/2} \quad . \tag{56}$$

A "smooth" trip will be associated with small σ, and a "rough" trip with a large one.

The concept of acceleration noise was introduced in conjunction with car-following studies (Montroll, 1961), and was further used to analyse the quality of driving under various conditions. For example, Jones and Potts (1962) used the concept in order to assess the safety and quality of driving along different types of highways (four-lane versus two-lane), the differences among drivers due to temperament and age, and the influence of the presence of attractions along the highway on the overall quality of driving. They reported the following conclusions:

- For two roads through hilly country, σ is much greater for a narrow two-lane road than for a four-lane divided highway.
- For a road in hilly country, σ is greater for a downhill journey than for an uphill one.
- For two drivers driving at different speeds below the design speed of a highway, σ is much the same.
- If one or both the drivers exceed the design speed, σ is greater for the faster driver.
- Increasing traffic volume increases σ.
- Increasing traffic congestion generated by parking cars, stopping buses, cross traffic, crossing pedestrians, etc., increases σ.
- The value of σ may be a better measure of traffic congestion than travel times or stopped times.
- High values of σ on a highway indicate a potentially dangerous situation.

Jones and Potts cautioned against the use of arbitrary interpretation of the values of σ, but they made the observation that $\sigma = 0.7$ ft/sec^2 is a low value, and $\sigma = 1.5$ ft/sec^2 is a high value.

Helly and Baker (1967) proposed another parameter, related to the acceleration noise, in order to describe the quality of driving, namely

$$G = \frac{\sigma}{u} \tag{57}$$

where G is the velocity gradient, σ the acceleration noise, and u the mean velocity during a trip. The reasoning behind their proposal of G as the measure of goodness was that acceleration noise alone does not describe the quality of the trip, because a fast trip and a slow trip could both have the same value of σ, but the fast trip would be more desirable. The time T used in their calculation of σ was the total trip time, including stopped time.

Underwood (1968) reported results of observations in Australia which suggest that G may be a better measure of traffic congestion than σ, but contradicts item 7 of the conclusions of Jones and Potts by questioning whether the velocity gradient is any better than total travel time as a measure of congestion.

As for the influence of car-following on acceleration noise, good measurements were reported by Herman and Rothery (1964) from actual experiments, shown in Fig. 14.

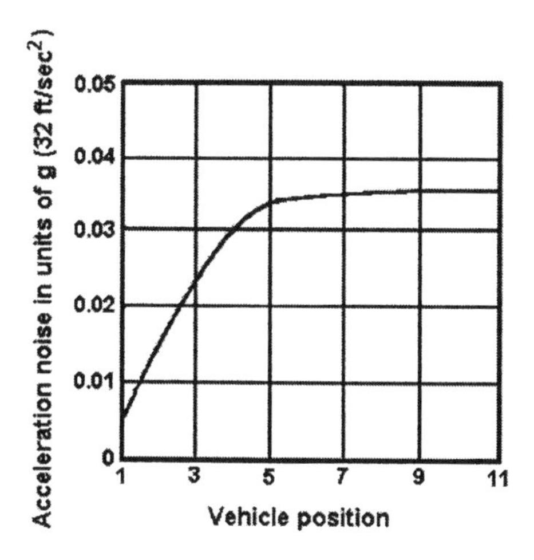

Figure 14. Acceleration noise in a platoon of cars

It is seen from this figure that the acceleration noise increases rapidly with the position of the car up to about the 5th position, after which it reaches a plateau. This result was obtained for an isolated platoon. In actual traffic, the movements of cars around the platoon will probably accentuate the disturbances to the platoon and the resulting acceleration noise.

1.3.4 Non-linear Car-Following Models, and Tie-in to Macroscopic Theory

The car-following model of traffic found a harmonious tie-in to macroscopic theory in a paper by Gazis, Herman, and Potts[23]. The bringing together of the two approaches proceeds as follows:

Let us assume that traffic consists of a line of cars following a leader driving at constant speed u_0. We then assume that the leader undergoes a manoeuvre, changing the car speed to a new speed u_1. Assuming that local and asymptotic stability are satisfied, we can integrate the basic car-following equation, Eq. 33, neglecting the time-lag *T*, since all cars will reach a new steady state after a sufficiently large time interval. In fact, we can re-write Eq. 33 in the form

$$\frac{du}{dt} = \lambda \frac{ds}{dt} \quad , \tag{58}$$

where u is the speed of the average vehicle and s is the spacing between vehicles. Equation 58 states that the acceleration of the average vehicle, du/dt, is proportional to the average relative speed, ds/dt. Integrating Eq. 58 we obtain

$$u = \lambda s + c_1$$

where c_1 is the integration constant. Its value can then be determined from the end condition that the speed be zero at the jam density s_j, hence $c_1 = -\lambda s_j$. Hence

$$u = \lambda(s - s_j) = \lambda\left(\frac{1}{k} - \frac{1}{k_j}\right)$$

$$q = ku = \lambda\left(1 - \frac{k}{k_j}\right)$$

and the flow versus concentration curve comes out as a straight line with a slope $-\lambda / k_j$. The intercept of this line with the q axis is the maximum value of flow, $q_m=\lambda$, when $u \to \infty$. This result does not agree with most observations, which indicate a flow versus concentration relationship represented by a convex curve, such as Greenberg's model shown in Fig. 6. But Gazis et al (1959) showed that the Greenberg steady-state model could be derived from car-following theory by a nonlinear model in which the sensitivity, or gain factor, was not a constant but was inversely proportional to the spacing between cars. In this case, the car-following equation becomes

$$\frac{dx_{n+1}(t+T)}{dt} = \lambda \frac{\left[\frac{dx_n(t)}{dt} - \frac{dx_{n+1}(t)}{dt}\right]}{[x_n(t) - x_{n+1}(t)]}$$

which, in the simplified form of Eq. 58, becomes

$$\frac{du}{dt} = \left(\frac{\lambda}{s}\right)\frac{ds}{dt} \tag{59}$$

Integrating Eq. 59, we obtain

$$u = \lambda \ln s + c_2 \tag{60}$$

The speed must be zero at jam density, hence $c_2 = -\lambda \ln s_j$, and Eq. 60 becomes

$$u = \lambda \ln \frac{s}{s_j} = \lambda \frac{k_j}{k} \quad ,$$

which is none other than Greenberg's flow versus concentration relationship. However, this model also suffers from the inappropriate limit for the velocity at zero concentration, since $u \to \infty$ as $k \to 0$. For this reason, Edie (1961) proposed a somewhat more complicated model which appeared to fit data for low concentrations better than the inverse spacing model. Edie's model, omitting time lags, is

$$\frac{du}{dt} = \left(\frac{\lambda u}{s^2}\right)\frac{ds}{dt} \quad , \tag{61}$$

which, when integrated, yields

$$k = k_m \ln\left(\frac{u_f}{u}\right) \tag{62}$$

where u_f is the "free-flow speed" at zero concentration. However, the disadvantage of the Edie model is that the concentration tends to infinity as the speed tends to zero.

All the above mentioned car-following models are included in a general non-linear model proposed by Gazis, Herman, and Rothery (1961), in which the sensitivity is a generalised function of the speed of the following car and the spacing between lead and following cars. The general form of this model is

$$\frac{d^2 x_{n+1}(t+T)}{dt^2} = \lambda \left(\frac{dx_{n+1}}{dt} \right)^m \frac{\left[\frac{dx_n(t)}{dt} - \frac{dx_{n+1}(t)}{dt} \right]}{\left[x_n(t) - x_{n+1}(t) \right]^l} \tag{63}$$

which, with appropriate choices for l and m, gives any of the above models, and more.

For data taken by car-following runs in the Lincoln tunnel of New York City, Gazis et al found the best correlation for values of $m = 1$ and $l = 2$. It is interesting to note that May and Keller (1969) investigated the possible choice of fractional exponents for freeway and tunnel data, and found that those giving the best correlations were $m = 0.8$, $l = 2.8$, for freeway data, and $m = 0.6$, $l = 2.1$, for tunnel data.

1.4 SOME QUESTIONABLE "PARADOXES"

As in many other fields of science, there have been some misguided contributions, some of which persisted in their popularity for a considerable time. One is reminded of Mark Twain's saying: "*There is something fascinating about science. One gets such wholesale returns of conjecture out of such a trifling investment of facts*". We shall discuss here two "paradoxes" based on traffic flow models, which are based on rather faulty reasoning, although they appear to be mathematically correct.

1.4.1 The Smeed paradox

In 1967, Smeed (1967) published a paper in which he discussed a most intriguing "paradox". The paper contained a mathematical proof, based on traffic flow models, that under certain circumstances a vehicle would reach its destination earlier by starting later. The Smeed paradox was based on the following argument. Suppose a car is standing at the entrance of a long section of the roadway, considering making an entrance toward its destination at the end of the roadway section.

The vehicle is supposed to be at the end of a platoon of n identical car, which in effect make a collective decision to enter at some chosen headway value. As the headway increases, the nth car has to leave later and later. However, larger headways mean lower densities, and according to flow theory higher speeds. Smeed went on to prove mathematically that under certain circumstances choosing a larger headway, and therefore delaying the departure, the car could arrive earlier than it would have with the unchanged headway.

The problem with Smeed's argument lies in the assumption that a vehicle entering a roadway section in effect influences the steady-state speed of vehicles in front. This amounts to an assumption that disturbances propagate *downstream*. However, as every driver knows, traffic is a uni-directional medium, in which disturbances can only propagate *upstream*. (Here, we can ignore the minor effect of a "tailgater" influencing a car in front to drive faster!) Another way of discerning the error in Smeed's reasoning is the following. Suppose the vehicle does not enter the roadway, but a "phantom" vehicle does, which does not influence the roadway density but moves with the steady-state speed of the traffic stream. After the vehicle enters, it will never be able to catch up with the phantom vehicle and overtake it, leading to the resounding conclusion that it must arrive later if it starts later. The conclusion is that Smeed, an outstanding leader in traffic science, slipped into an erroneous line of reasoning in this case.

The erroneous consideration of forward propagation of disturbances is not limited to the Smeed paradox. In recent years, traffic scientists with an electrical Engineering background started talking about roadways as having *impedance* properties. The idea is that just as wires have their resistance increased with increasing electric current flow, highways have their impedance increased, slowing down the traffic movement, when the volumes of traffic are increased, or when the density of traffic is increased. This is all fine, except that cars do not transmit information from one to another as fast as electrons, and the signals they transmit are unidirectional, going upstream. This means that the concept of impedance of roadways can only be applied for steady state conditions of traffic movement. Applying it to dynamic traffic phenomena runs the danger of fallacy similar to that of the Smeed paradox.

1.4.2 The Braess paradox

The Smeed paradox never gained wide acceptance, because several traffic theorists, including the author of this book, saw its fallacy and communicated it widely. However, there is another paradox, the Braess paradox (1968), which has received wide publicity and continues to be accepted as the absolute truth. The Braess paradox states that building an additional link in a network may not only fail to improve the performance of the network, but it may decrease it. The general idea of the paradox describes the performance of a network such as shown in Fig. 15. It is assumed that cars enter at point A and exit at point B. One starts with a network that contains only the links AB, AC, BD, and CD, and asks whether or not it makes sense to build the link BC. Braess assumed some "performance" value for each of the four existing links as a function of the flow along the link, which corresponded to the delay to the traffic stream in the link. He computed the delay for the entire network by assuming that

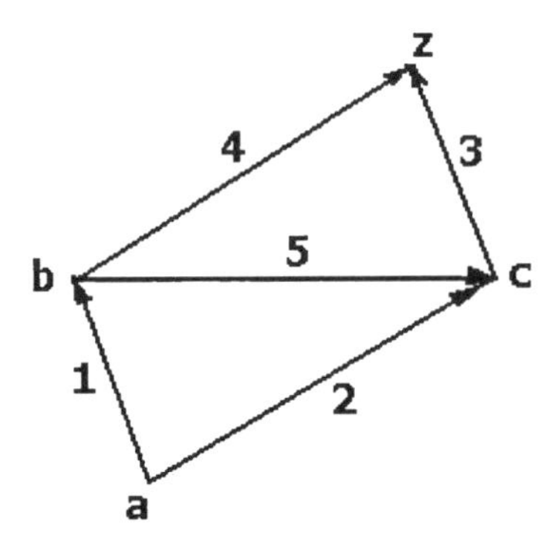

Figure 15. The Braess Paradox

the stream splits among the two routes, abz and acz. Assuming congested conditions, any increase of the flow along any link would substantially increase the delay to drivers. Braess went on to compute the performance of the network after addition of the link bc, which would attract some of the traffic along abz. He found that under certain circumstances the penalty to the cars would increase in the new network, because even an improvement along bz would be far exceeded by the worsening along cz.

The problem with the Braess paradox is that one can question the specific delay functions that he assumed. In the example he gave corresponding to Fig. 15, these delay functions, if assumed as multiples of the inverse of speed, would correspond to extraordinary, and very debatable, flow versus concentration relationships. For example, he assumed for link 1 the delay function $10q$, where q is the volume. If we set the delay as l_1/v, where l_1 is the link length, then we obtain the corresponding q versus k (density) relationship $q = (kl_1/10)^{1/2}$, a rather unlikely relationship for which q tends to infinity as k goes to infinity.The distribution of traffic after the addition of the link is obtained from an optimisation procedure, which depends on these unrealistic delay functions. For an uncongested system, anyone familiar with network flow theory recognises that the "minimal cut" of the network cannot possibly get worse after addition of a link. For a congested system, the point z could only be viewed as an exit point, which will probably be the bottleneck, and its performance is not in any way affected by the addition of a link. The most that can be said in this case is that the addition of the link does not improve the performance of the system, but it does not make it worse either. To be sure, the addition of an attractive, new link may draw some drivers into a route choice which degrades the performance of the system, even for realistic choices of delay functions. This does not mean that the system has worsened, and it may very well be that trial and error, plus some sensible instructions to drivers, can bring drivers into a better set of route choices. In any case, the conflict between individual route choices in a network and global optimisation objectives has been well known for many decades, and there is nothing paradoxical about it, as discussed in Chapter 4 of this book.

There is even an element of forward propagation of traffic noise in the Braess discussion. A diversion of traffic along the link 5 (Fig. 15) is assumed to change the behaviour of the forward link 3 only. In real life, it is likely to affect even more the link 2, particularly during periods of congestion, when the node C may become a "Throughput Limiting Point (TLP), as discussed in Chapter 3 of this book.

So, the Braess paradox is an attractive statement, which has even generated articles in the popular press, but it not a very helpful design tool for transportation planners. In recent years, many people have cited the Braess paradox as correct, in the sense that construction of new roadways, or improvement of existing ones, attracts new traffic, which results in worse traffic conditions than before construction or improvement. This "traffic generation" effect may be true, but it has nothing to do with the original statement of the Braess paradox.

1.5 ADDITIONAL FLOW THEORY WORK, AND FUTURE CHALLENGES

Several models of traffic systems and traffic flow have been proposed in recent years, including those based on "chaos theory", "cellular automata", and "hopping models". Some of them had the advantage of providing a convenient simulation tool, and in this way may have been helpful to researchers, since calibration of these models showed them to be reasonably good in describing actual traffic conditions. However, many of these models fail to describe the actual behaviour of traffic, and the partial fit with some measurement does not justify their promotion as improved modelling tools. For example, the hopping models essentially convert automobile traffic into what might be called a "moving block" system, using the nomenclature of railway traffic, and it is very unlikely that this is how people drive. The chaos theories, while very attractive to journalists who love the concept of chaotic behaviour of traffic, do not contribute any understanding of this behaviour. Many of the seemingly chaotic features of traffic can be explained as the effects of accidents, the geometry of the roadways, or such phenomena as the "moving and phantom bottleneck" discussed earlier in this chapter.

The above comments do not mean to imply that the modelling of traffic flow is finished. There are many challenges ahead, and meeting these challenges holds promise of additional tools for improved traffic management. One of the needed key improvements of traffic flow theory concerns the effect of roadway geometry.

Virtually all the traffic flow models discussed in this chapter are good mainly for an infinitely long, flat, and straight roadway. However, every good driver knows that the geometry of the roadway affects driving patterns. For example, a curve may necessitate some slowing down of traffic before the curve with some acceleration along the curve and past the curve. This effect would alter the nature of either the macroscopic or the microscopic, car-following models of traffic. An additional effect of geometry is that due to inclined roadway sections. An example of the effect of such sections was given in this chapter, in the discussion of the "moving and phantom bottleneck", which may result from the slowing down of a truck by an uphill portion of the roadway. It should also be mentioned that the effect of uphill portions has been considered in the case of tunnel traffic, discussed in Chapter 3 of this book. In general, work is needed in augmenting traffic flow models to include the effects of geometry. Such augmentation will be particularly needed if automation ever becomes a standard replacement of drivers, or if railroad train control progresses to “moving block”, automated train control.

References

Anderson, R. L., Herman, R. and Prigogine, I. (1962), "On the statistical distribution function theory of traffic flow", *J. Opns. Res.*, **10**, 180-195.

Braess, D. (1968), "Uber ein Paradoxen der Verkehrsplannung", *Unternehmenstorchung,* **12**, 258-268.

Chandler, R. E., Herman, R. and Montroll, E. W. (1958), "Traffic Dynamics: Studies in Car Following", *Operations Res.*, **6**, 165-184.

Daou, A. (1964), "On Flow in Platoons", *Bull. Operations Res. Oc. Amer.*, 25ty National Meeting, **12,** B40.

Edie, L. C. (1961), "Car-Following and Steady-State Theory for Non-Congested Traffic", *Operations Res.*, **9**, (1), 66-76.

Edie, L. C. (1974), "Flow Theories", in *Traffic Science*, edited by D. C. Gazis, Wiley-Interscience, New York, pp. 1-108.

Forbes, T. W., Zagorski, M. J., Holshouser, E. L. and Deterline, W. A. (1958), "Measurement of Driver Reactions to Tunnel Conditions", *Proc. Highway Res. Board,* **37,** 345-357.

Gazis, D. C., Herman, R. and Potts, R. B. (1959), "Car-Following Theory of Steady-State Traffic Flow", *Operations Res.*, **7**, 499-505.

Gazis, D. C., Herman, R. and Rothery, R. W. (1961), "Nonlinear Follow-the-Leader Models of Traffic Flow", *Operations Res.*, **9**, 545-567.

Gazis, D. C., Herman, R. and Rothery, R. W. (1963), "Analytical Methods in Transportation: Mathematical Car-Following Theory of Traffic Flow", *Proc. Am. Soc. Civil Engrs, Eng. Mech. Div.,* **6**, 29-46.

Gazis, D. C. and Herman, R. (1992), "The Moving and 'Phantom' Bottlenecks", *Transp. Science*, **26,** No.3, 223-229.

Greenberg, H. (1959), "An Analysis of Traffic Flow", *Operations Res.,* 7, No.1, 79-85.

Helly, W. and Baker, P. G. (1967), "Acceleration noise in a congested signalized environment", in *Vehicular Traffic Science, Proc. Third Intern. Symp. on Th. Of Traffic Flow,* American Elsevier, New York, pp. 56-61.

Herman, R., Montroll, E. W., Potts, R. B. and Rothery, R. W. (1959), "Traffic Dynamics: Analysis of Stability in Car Following", *Operations Res.,* **7,** 86-106.

Herman, R. and Potts, R. B. (1961), "Single-Lane Traffic Theory and Experiment", in *Theory of Traffic Flow,* edited by R. Herman, Amsterdam:Elsevier, 120-146.

Herman, R. and Rothery, R. W. (1962), "Microscopic and Macroscopic Aspects of Single Lane Traffic Flow", *Operations Res. Japan*, 5, 74.

Herman, R. and Prigogine, I. (1979), "A Two-Fluid Approach to Town Traffic", *Science,* **204**, 148-151.

Jones, T. R. and Potts, R. B. (1962), "The measurement of acceleration noise – a traffic parameter", *Oper. Res.* **10** (6), 745-763.

Lighthill, M. J. and Whitham, G. B. (1955), "On Kinematic Waves, II. A Theory of Traffic Flow on Long Crowded Roads", *Proc. Roy. Soc. (London)* **229A,** 317-345.

Mahmassani, H. S., Williams, J. C. and Herman, R. (1987), "Performance of Urban Traffic Networks", in *Transportation and Traffic Theory,* edited by N. H. Gartner and N. H. M. Wilson, Elsevier Science Publishing Company, Amsterdam, pp. 1-20.

May, A. D. and Keller, H. E. M. (1969), "Evaluation of Single and Multi-Regime Traffic Flow Models", in *Beitrage zur Theorie des Verkehrs Flusses,* Bonn: Heransgegeben vom Bundesminister fur Verkehr, pp. 37-47.

Montroll, E. W. (1961), "Acceleration noise and clustering tendency of vehicular traffic", in *Proc. Symp. on Th. Of Traffic Flow*, Elsevier, New York, pp. 147-157.

Newell, G. F. (1965), "Instability in Dense Highway Traffic", in *Proc. 2nd Intrn. Symp. on the Theory of Road Traffic Flow*, edited by J. Almond, Paris: Office of Economic Cooperation and Development, 73-83.

Pipes, L. A. (1953), "An Operational Analysis of Traffic Dynamics", *J. Appl. Phys.*, **24**, 274-281.

Prigogine, I. (1961), "A Boltzmann-like approach to the statistical theory of traffic flow", in *Theory of Traffic Flow,* R. Herman, Editor, Elsevier Publishing Co., Amsterdam.

Reuschel, R. (1950), "Fahrzeughewegungen in der Kolonne bei gleichformig beschleunigtem oder verzogertem Leitfarzeug", *Zeit. Osterr. Ing. und Arch. Ver.,* **95,** 52-62, 73-77.

Richards, P. I. (1956), "Shock Waves on a Highway", *Operations Res.,* **4**, 42-51.

Smeed, R. J. (1967), "Some Circumstances in Which Vehicles Will Reach Their Destination Earlier by Starting Later", *Transp. Sci.*, **1**, 308-317 .

Underwood, R. T. (1968), "Acceleration noise and traffic congestion", *Traffic Eng. Control*, **10** (3), 120-123, 130.

Additional References

Buckley, D. J. (1962), "Road Traffic Headway Distributions", *Proc. Aust. Road Res. Board*, **1**, 153-187.

Daganzo, C. F. (1997), *Fundamentals of Transportation and Traffic Operations*, New York: Elsevier.

Darroch, J. N. and Rothery, R. W. (1972), "Car Following and Spectral Analysis", in *Traffic Flow and Transportation*, edited by G. F. Newell, New York: Elsevier.

Dunne, M. C., Rothery, R. W. and Potts, R. B. (1968), "A Discrete Markov Model of Vehicular Traffic", *Transp. Sci.*, **2**, No. 3, 233-251.

Gazis, D. C., Herman, R. and Weiss, G. (1962), "Density Oscillations Between Lanes of a Multi-Lane Highway", *Operations Res.,* **10**, 658-667.

Greenberg, H. (1959), "An Analysis of Traffic Flow", *Operations Res.*, **7**, No. 1, 79-85 (1959).

Greenshields, B. and Loutzenheiser, A. (1940), "Speed Distributions", *Proc. Highway Res. Board*, **20**.

Haight, F. A. (1963), *Mathematical Theories of Traffic Flow*, New York: Academic Press.

Kometani, E. (1955), "On the Theoretical Solution of Highway Traffic Under Mixed Traffic", *Jour. Faculty, Kyoto Univ.*, **17**, 79-88.

Miller, A. J. (1961), "Analysis of Bunching in Rural Two-Lane Traffic", *J. Roy. Statist. Soc.*, **B-23**, 64-75.

Montroll, E. W. and Potts, R. B. (1964), "Car-Following and Acceleration Noise", *An Introduction to Traffic Flow Theory*, Highway Research Board Special Report 79.

Newell, G. F. (1955), "Mathematical Models of Freely Flowing Highway Traffic", *Operations Res.*, **3**, 176-186.

Newell, G. F. (1962), "Theories of Instability in Dense Highway Traffic", *J. Oper. Res. Soc. Japan*, **5**, 9-54.

Newell, G. F. (1965), "Instability in Dense Highway Traffic", in *Proc. 2nd Intern. Symp. on the Theory of Road Traffic Flow*, edited by J. Almond, Paris: Office of Economic Cooperation and Development, pp. 73-83.

Newell, G. F. (1968), "Stochastic Properties of Peak Short-Time Traffic Counts", *Transp. Sci.*, **2**, No. 3, 167-183.

Oliver, R. M. (1952), "Distributon of Gaps and Blocks in a Traffic Stream", *Oper. Res.*, **10**, 197-217.

Prigogine, I., Herman, R. and Anderson, R. (1965), "Further Developments in the Boltzmann-like Theory of Traffic", in *Proc. 2nd Intern. Symp. on the Theory of Road Traffic Flow*, edited by J. Almond, Paris: Office of Economic Cooperation and Development, pp. 129-138.

Prigogine, I. and Herman, R. (1971), *Kinetic Theory of Vehicular Traffic*, New York: Elsevier.

Underwood, R. T. (1963), "Traffic Flow and Bunching", *J. Aust. Rd. Res.*, **1**, 8-25.

Vaughn, R. and Hurdle, V. F. (1992), "A Theory of Traffic Flow for Congested Conditions on Urban Arterial Streets. *I*: Theoretical Development", *Transp. Res.*, **26B**, 381-396.

Wardrop, J. G. (1952), "Some Theoretical Aspects of Road Traffic Research", *Proc. Inst. Civil Engineers*, Part II, **1**, No. 2, 325-362.

Wright, C. C. (1972), "Some Properties of the Fundamental Relations of Traffic Flow", in *Traffic Flow and Transportation*, edited by G. F. Newell, New York: Elsevier, pp. 19-32.

Chapter 2

Queueing and Delays at Isolated Intersections

A driver is delayed along his journey by a combination of factors, which includes the interactions with other cars and the effect of regulatory devices such as stop signs and traffic signals. In order to improve transportation systems, it is desirable to be able to estimate the effects of any regulatory devices which are to be used. Such estimates can be achieved through the use of the methodologies of queueing theory, which was originally developed for the study of communication systems. To be sure, many of the standard techniques of queueing theory, such as the assumption of specific functional character of time intervals between events, are only approximations to the values encountered in real systems. Nevertheless, the study of traffic system behavior through the methods of queueing theory gives us a handle for designing improvements in such systems.

In this chapter, we discuss the effect of isolated intersections on the delay to drivers. In Chapter 3, we shall discuss the effect of coordination of a system of traffic signals in reducing such delays. In this chapter, we discuss the effect of stop or yield signs, as well as the effect of traffic signals. The estimates are derived through the use of the theory of stochastic processes and, in particular, on the methodologies of renewal processes and queueing theory.

2.1 TRAFFIC CHARACTERISTICS

Traffic moving along a highway can be described as a function of space, and as a function of time. For the purpose of the present discussion, it is particularly useful to consider the intervals of time between events, such as the arrival times of individual vehicles at a particular point. Let the sequence of arrival times, at a fixed point, of successive cars in an unimpeded stream of traffic be $t_1, t_2, t_3, \ldots$. The *headways* between cars, G_r, are defined as

$$G_r = t_r - t_{r-1}, \qquad r = 2,3,4,\ldots$$

and G_1 is defined to be equal to t_1 whether or not the time $t = 0$ corresponds to the arrival of a car at the fixed point. The assumption made in most traffic studies is that the successive headways are independent and identically distributed random variables. Let us denote the probability function for the successive gaps G_i by $\varphi(G)$. If the instant $t = 0$ at which measurements are begun is chosen at random, uncorrelated with the arrival of any individual car, then the probability density function for G_1, denoted by $\varphi_0(G)$, is given by (see Adams, 1936)

$$\varphi_0(G) = \frac{1}{\mu}\int_0^{\infty} \varphi(x)dx \tag{1}$$

where μ is the mean headway given by

$$\mu = \int_0^{\infty} G\varphi(G)dG \tag{2}$$

It is expected that, in any realistic model of traffic, μ is finite. The most widely used form for $\varphi(G)$ in traffic studies is the negative exponential

$$\varphi(G) = \varphi_0(G) = \frac{1}{\mu}\exp\left(-\frac{G}{\mu}\right) \tag{3}$$

This density function was first observed experimentally by Adams. Later, theoretical justification for this choice was given by Weiss and Herman (1962), and also by Breiman (1963) and Thedeen (1964). Weiss and Herman derive their proof in the following way.

Consider a stream of traffic on an infinitely long, homogeneous highway. Suppose that each car is represented by a dimensionless point traveling at constant speed, υ, which is sampled from a probability density function $f(\upsilon)$, which is not a delta function or a combination of delta functions. We also assume that if a car reaches a slower car, it can pass immediately without incurring any delay. Let us further assume that at time $t = 0$ the cars are distributed on the highway in a way such that the probability that a car is located between x and $x + dx$ along the highway is $\rho_0(x)dx$, where $\lim_{x\to\pm\infty} \rho_0(x) = \rho_0 =$ constant. This would be the case, for example, if successive headways were identically distributed random variables with a finite second moment. The preceding assumptions fully define the dynamics of changes in the headways of the cars, and so we can calculate the probability, $\rho(x,t)dx$, that at time t, the distance between a given car and any other car will be between x and $x + dx$. A detailed consideration of the movement of the two cars, at speeds υ and υ', yields

$$\rho(x,t) = \int_0^\infty \int_0^\infty \rho_0[x + (\upsilon - \upsilon')t] f(\upsilon) f(\upsilon') d\upsilon d\upsilon' \tag{4}$$

For $t \to \infty$, this becomes

$$\lim_{t\to\infty} \rho(x,t) = \rho_0 \int_0^\infty \int_0^\infty f(\upsilon) f(\upsilon') d\upsilon d\upsilon' = \rho_0 \tag{5}$$

Thus $\rho(x,\infty)$ is a constant. Translating this result into the time domain, we conclude that the probability density for the passage of two successive cars past a fixed point is

$$\varphi(t) = q \exp(-qt) \ , \tag{6}$$

where q is the mean flow given by

$$q = \rho_0 \int_0^\infty \upsilon f(\upsilon) d\upsilon \quad . \tag{7}$$

The preceding discussion does not take into account explicitly driver and vehicular interactions, which would alter the resulting headway distribution. There have been numerous contributions toward making the headway distributions more realistic than the idealized exponential distribution. Among other issues, the exponential distribution assigns a relatively high weight to very small gaps, near zero, which are in effect impossible due to the finite size of cars. One suggestion was that of Schuhl (1955) who suggested the function

$$\begin{aligned} \varphi(t) &= 0 \qquad t \le T \\ \varphi(t) &= q \exp[-q(t-T)] \ , \ t > T \end{aligned} \tag{8}$$

Obviously, the introduction of additional parameters increases the likelihood of a good fit to observations. However, for the purposes of our discussion, the exponential distribution is sufficiently successful in giving some information regarding the queueing phenomena in traffic.

There is another important contribution to the study of successive headways between cars, which should be mentioned here, and it is the description of traffic as an aggregate of "traveling queues", by Miller (1961). He rightly observed that because of restrictions in passing capability, cars tend to bunch into platoons which move at the speed of a leader, leaving gaps between successive bunches. The definition of a bunch, or queue, is arbitrary, but Miller defined it by the requirement that headways between successive cars be no more than 8 sec., and the relative speed of cars within a queue be within the range of -3 to 6 miles/hr. Successive queues were assumed to be independent, and the spacing between queues was assumed to follow a negative exponential distribution. Miller reported good agreement with data provided by the Swedish State Roads Institute. The distribution of the number of cars in each queue was found to be described equally well by the distributions

$$p_n = \frac{n^{n-1}}{n!} r^{n-1} e^{-r^n}$$
$$p_n = \frac{(m+1)(m+1)!(n-1)!}{(m+n-1)!} \tag{9}$$

where n is the number of cars in the queue, and r and m are parameters fitted from the data.

2.2 THE GAP ACCEPTANCE FUNCTION

A driver or a pedestrian, wishing to move forward, is often impeded by traffic which is obstructing his path. To describe this situation, the concept of "gap acceptance" has been introduced. The idea is that a driver does not move until the headway between successive cars blocking his path, otherwise known as the "gap", exceeds a certain critical value. If this gap is denoted by G, then the behavior of the driver is summarized by a gap acceptance function $\alpha(G)$, which represents the probability that the driver will move when facing a gap G. The simplest form of the gap acceptance function is a step function

$$\alpha(G) = H(G-T) , \tag{10}$$

where $H(x)$ is the delta function; namely, $H(x)=0$ for $x<0$, and $H(x)=1$ for $x \geq 0$. Measurements of actual driver behavior reveal a more complicated behavior than that described by Eq. 10. For this reason, alternative expressions for $\alpha(G)$ have been suggested, in conjunction with experimental evidence. For example, Herman and Weiss (1961) found that, on the basis of experiments conducted under somewhat idealized conditions, the gap acceptance function could be represented by

$$\alpha(G) = 0 , \quad G < T$$
$$\alpha(G) = 1 - \exp[-\lambda(G-T)] , \quad G \geq T \tag{11}$$

where the parameters λ and T were found to be $\lambda = 2.7$ sec^{-1} and $T = 3.3$ sec.

In addition to the gap acceptance function by an individual, the "population gap acceptance function", α_p , has been defined as the probability that a gap will be accepted by a randomly selected member of a population of drivers. Assuming that $\alpha(G)$ for an individual driver is a step function $H(G-T)$, where T is a random variable with a probability density $u(T)$, the population gap acceptance function is

$$\alpha_p(G) = \int_0^\infty H(G-T)u(T)dT = \int_0^G u(T)dT \qquad (12)$$

The population gap acceptance function is useful in interpreting field data on gap acceptance, which by necessity involve the compilation of acceptable gaps by a population of drivers, rather than experiments with a single driver.

At this point, some remarks are appropriate regarding the realism of the presented models of gap acceptance. One obvious observation is that human behavior may not be as simplistic as stipulated by the models which assume that the size of the acceptable gap remains constant for a given individual. A few observations will convince anyone that this may not be the case, because the patience of a driver may be tested to the point of shrinking the size of acceptable gaps. Another effect on gap acceptance is the succession of gaps. A driver looking for an opening may skip a gap which appears to be acceptable, if he detects an even more pleasing gap coming next. Nevertheless, the methodology presented here gives a handle for estimating the delays to traffic movement caused by traffic competing for the same roadway space, for the purpose of planning improvements to transportation systems.

2.3 THE DELAY TO A SINGLE CAR

Let us now assume that a single car arrives at a stop sign at time $t = 0$, and faces an unobstructed stream of traffic while trying to either cross, or merge with the main stream of traffic. We assume that the driver's gap acceptance function is $\alpha(t)$. We will calculate the probability density , $\Omega(t)$, of the delay to the driver, following the work of Weiss and Maradudin

(1962). In what follows, we will talk about a "merging maneuver", but the results are also applicable to the case of crossing a single stream of main traffic.

The delay time is either zero, if an acceptable gap is found at $t = 0$, or it is a positive quantity if one or more cars pass the stop sign and the merging maneuver takes place immediately after the passage of a car on the main road. Let us assume that successive headways along the main road are identically distributed independent random variables with probabilitydensity $\varphi(t)$. The probability of an initial acceptable gap, denoted by α_0, is given by

$$\bar{\alpha}_0 = \int_0^\infty \alpha(t)\varphi_0(t)dt \tag{13}$$

and therefore we can express $\Omega(t)$ in the form

$$\Omega(t) = \bar{\alpha}_0\,\delta(t) + \Omega_1(t) \tag{14}$$

where $\delta(t)$ is the Dirac delta function, and $\Omega_1(t)$ represents the contribution due to a delayed merge. The probability that an arbitrary gap is acceptable is $\bar{\alpha} = \int_0^\infty \alpha(t)\varphi(t)dt$, and therefore

$$\Omega_1(t) = \bar{\alpha}w(t) \tag{15}$$

where $w(t)dt$ is the probability that a car on the main road passes the intersection in the time interval $(t, t + dt)$, and no merge has been made up to that point. We now show that $w(t)$ is the solution to an integral equation of convolution type. Let us define two functions

$$\begin{aligned} \Psi_0(t) &= \varphi_0(t)[1 - \alpha(t)] \\ \Psi(t) &= \varphi(t)[1 - \alpha(t)] \end{aligned} \tag{16}$$

which have the following meaning: $\Psi_0(t)\,dt$ is the probability that the first gap is between t and $t + dt$ and is unacceptable, and $\Psi(t)\,dt$ has the same meaning for succeeding gaps. If a car in the main stream passes the intersection at time t and no merge has been made, it is either the first car to do so, or the last such event occurred at some time $\tau < t$ and the succeeding gap was unacceptable. These two possibilities are summed up in the equation

$$w(t) = \Psi_0(t) + \int_0 w(t)\Psi(1-\tau)d\tau \tag{17}$$

Since Eq. 17 contains a convolution integral, the use of Laplace transforms is appropriate. Let us denote the Laplace transform of a function of t by the same symbol of a function of s with an asterisk, so that

$$\begin{aligned}\Omega^*(s) &= \int_0^\infty e^{-st}\Omega(t)dt \\ \Psi^*(s) &= \int_0^\infty e^{-st}\Psi(t)dt\end{aligned} \tag{18}$$

The Laplace transform of Eq. 17 leads to the relationship

$$w^*(s) = \frac{\Psi_0^*(s)}{1-\Psi^*(s)} \tag{19}$$

Equations 14, 15, and 19, in concert, yield

$$\Omega^*(s) = \overline{\alpha}_0 + \frac{\overline{\alpha}\Psi_0^*(s)}{1-\Psi^*(s)} \tag{20}$$

It may be possible, in some cases, to obtain an explicit expression for $\Omega(s)$ by inverting Eq. 20. However, in most cases, it is sufficient to use Eq. 20

as a moment generating function, and obtain the moments of delay time. To this end, we use the relationship for the moments

$$\overline{t^n} = \int_0^\infty t^n \Omega(t)dt = (-1)^n \frac{d^n}{ds^n}\Omega^*(s)\,|_{s=0} \tag{21}$$

Using this formula, we find that the first two moments of the delay are given by

$$\begin{aligned}\bar{t} &= \int_0^\infty t\left[\Psi_0(t) + \left(\frac{1-\bar{\alpha}}{\bar{\alpha}}\right)\Psi(t)\right]dt \\ \overline{t^2} &= \int_0^\infty t^2\left[\Psi_0(t) + \left(\frac{1-\bar{\alpha}_0}{\bar{\alpha}}\right)\Psi(t)\right]dt + \frac{2\bar{t}}{\bar{\alpha}}\int_0^\infty x\Psi(x)dx\end{aligned} \tag{22}$$

When $\varphi\ (t)\ =\ (\ 1/\mu\ )\ \exp\ (-t/\mu)$ and $\alpha\ (t)$ is the step function $\alpha(t) = \mathrm{H}(t-T)$, the expression for $\Omega^*(s)$ becomes

$$\Omega^*(s) = e^{-T/\mu}\frac{s+(1/\mu)}{s+(1/\mu)\exp\{-[s+(1/\mu)T]\}} \tag{23}$$

If we now expand the right-hand side of Eq. 23 in series, and invert term by term, we obtain

$$\Omega(t) = e^{-t/\mu}\delta(t) + e^{-T/\mu}\sum_{n=1}^{\infty}\frac{(-1)^n}{(n-1)!\,\mu^n}\{e^{-nT/\mu}(1-nT)^{n-1}H(1-nT)$$

$$- e^{-(n-1)T/\mu}[t-(n-1)T]^{n-1}H[t-(n-1)]\ \} \tag{24}$$

The mean and variance of the delay time are readily obtained from the expression for $\Omega^*(s)$ in Eq. 23, and they are

$$\bar{t} = \mu\left(e^{T/\mu} - 1 - \frac{T}{\mu}\right)$$
$$\sigma^2 = \mu^2\left[e^{2T/\mu} - \left(\frac{2T}{\mu}\right)e^{T/\mu} - 1\right] \tag{25}$$

Some interesting variations on the preceding discussion have been worked out to account for variations in the behavior of the driver attempting to merge. The first one addresses the difference dictated by a yield sign as opposed to a stop sign. In this case, a driver reaching the intersection at some speed may be willing to accept a smaller first gap than the subsequent gaps, which he has to face after coming to a complete stop. If we assume that in this case $\alpha_0(t)$ is the gap acceptance function for the first gap, and $\alpha(t)$ that for succeeding gaps, then in Eqs. 17 and 22, α_0 and $\Psi_0(t)$ are to be replaced by

$$\overline{\alpha}_0 = \int_0^\infty \alpha_0(t)\varphi_0(t)dt$$
$$\Psi_0(t) = \varphi_0(t)[1 - \alpha_0(t)] \tag{26}$$

If we further assume

$$\alpha_0(t) = \mathrm{H}(t - T_0)$$
$$\alpha(t) = \mathrm{H}(t - T) \tag{27}$$
$$\varphi(t) = \frac{1}{\mu}\exp\left(-\frac{1}{\mu}\right)$$

then the first and second moments of delay are found to be

$$\bar{t} = \mu\left(e^{T/\mu} - e^{(T-T_0)/\mu} - \frac{T}{\mu} + \frac{(T-T_0)}{\mu}e^{-T_0/\mu}\right)$$
$$\sigma^2 = \mu^2\left\{\left(e^{T/\mu} - \frac{T}{\mu}\right)^2\left(1 - e^{-2T_0/\mu}\right) - \frac{2T_0}{\mu}\left(e^{T/\mu} - \frac{T}{\mu}\right)e^{-2T_0/\mu} - \frac{T_0^2}{\mu^2}e^{-T_0/\mu}\left(1 + e^{-T_0/\mu}\right) - \frac{T^2}{\mu^2}e^{-T/\mu}\left(1 - e^{-T_0/\mu}\right)\right\} \quad (28)$$

Two variations on the preceding treatment have been considered by various researchers. Weiss and Maradudin (1962) considered the case of the "impatient driver", whose gap acceptance behavior changes with passing time, so that

$$\alpha_1(t) \le \alpha_2(t) \le \alpha_3(t) \le \ldots\ldots\ldots \quad (29)$$

where $\alpha_j(t)$ is the gap acceptance function for the gap preceding the arrival of the jth car on the main highway. Drew et al (1967a, 1967b) also considered this case and gave detailed results for the case in which the gap acceptance is a step function with a threshold decreasing in some regular manner with successive passages of cars on the main highway.

The second variation concerns a driver who considers more than the gap he faces at any moment. Specifically, it considers a driver who also looks at the next following gap, and waits for that one if it is larger that the one just in front even though it is acceptable. This case was considered by Weiss (1966), who treated the case of a negative exponential headway distribution and a certain probability that the driver would skip an acceptable gap for the sake of a larger following gap. The probability function assumed by Weiss was

$$p(G) = \exp[-\rho(G-T)]H(G-T) \quad (30)$$

where T is the critical gap and ρ is a parameter associated with the probability that a larger following gap is preferred by the driver. It is interesting to note that numerical results obtained on the basis of Eq. 30 show that the added delay due to such behavior on the part of the driver are negligible compared to the basic delay due to gap acceptance limitations.

Another interesting generalization of the treatment, treated by Weiss (1967), concerns the case in which the gap acceptance function depends not only on the time gap, but also on the speed of the oncoming vehicle on the main road. In this case, let us denote the gap acceptance function by $\alpha(\upsilon,t)$, and let $\varphi(\upsilon,t)d\upsilon dt$ be the probability that the gap is between t and $t+dt$, and the speed of the oncoming vehicle is between υ and $\upsilon + d\upsilon$. The treatment given for the standard case can be adapted by making the substitutions for a few key variables

$$\varphi_0(\upsilon,t) = \frac{\int_t^\infty \varphi(\upsilon,\tau)d\tau}{\int_0^\infty dt \int_0^\infty t\varphi(\upsilon',t)d\upsilon'}$$

$$\bar{\alpha} = \int_0^\infty \int_0^\infty \varphi(\upsilon,t)\alpha(\upsilon,t)d\upsilon dt \tag{31}$$

$$\Psi(t) = \int_0^\infty \varphi(\upsilon,t)[1-\alpha(\upsilon,t)]d\upsilon$$

and analogous substitutions for the variables α_0 and $\Psi_0(t)$. Detailed calculations were made for

$$\varphi(\upsilon,t) = \varphi_0(\upsilon,t) = \frac{f(\upsilon)}{\mu}\exp\left(-\frac{1}{\mu}\right) \tag{32}$$

which corresponds to a speed-headway distribution for light traffic. If, in addition, $\alpha(\upsilon,t)$ can be expressed as

$$\alpha(\upsilon,t) = H[t - T(\upsilon)] \tag{33}$$

then the mean delay is given by

$$\bar{t} = \frac{\mu}{\bar{\alpha}}\left\{1 - \int_0^\infty [1+\lambda T(\upsilon)]f(\upsilon)e^{-T(\upsilon)/\mu}d\upsilon\right\} \tag{34}$$

If $T(\upsilon)$ is a linear function of υ,

$$T(\upsilon) = \frac{T_0\upsilon}{\bar{\upsilon}} \tag{35}$$

where υ is the mean speed, and $f(\upsilon)$ is

$$f(\upsilon)=\frac{a^{r+1}\upsilon^{r}}{\Gamma(r+1)}e^{-a\upsilon} \tag{36}$$

with a and r representing the mean and variance of $f(\upsilon)$, namely,

$$\begin{aligned} a &= \frac{\bar{\upsilon}}{\sigma^2} \\ r &= \left(\frac{\bar{\upsilon}}{\sigma}\right)^2 - 1 \end{aligned} \tag{37}$$

The mean delay comes out to be

$$\frac{\bar{t}}{\mu}=\frac{1}{\bar{\alpha}}-1-\frac{T_0\bar{\upsilon}^2}{\mu\bar{\upsilon}^2+T_0\sigma^2} \tag{38}$$

where

$$\bar{\alpha}=\left(1+\frac{T_0\sigma^2}{\mu\bar{\upsilon}^2}\right)^{-(\bar{\upsilon}/\sigma)^2} \tag{39}$$

Numerical results presented by Weiss (1967) seem to indicate that the effect of the dependence of gap acceptance on the speed of the oncoming vehicle is not very pronounced, and they are unlikely to produce measurable changes in the merging delay.

There is an interesting departure from the treatment of the intersection delay problem in term of finding an acceptable gap among the headways of successive individual cars. Instead, cars are considered as arriving in bunches, or platoons. Miller (1961), who did considerable work in modeling platoons, proposed this "random queue" model of gap acceptance. A queue, or platoon, was defined as a group of cars whose speed and relative spacing did not vary more than a certain small amount, and it was assumed that no merge was possible while a queue was passing the merge point. Let the probability density function (p.d.f.) of the time of passage of a queue be $q(t)$, and let the p.d.f. of the headway between queues, i.e. the difference in time between the end of one queue and the beginning of the next, be $\varphi(t)$. In addition, let us define the mean duration of a queue as $v = \int_0^\infty tq(t)dt$, and the mean associated with $\varphi(t)$ as μ. Then, by arguments based on Renewal Theory (Cox, 1962), the probability that the merging car arrives during an interval between two successive queues is

$$p = \frac{\mu}{\mu + v} \tag{40}$$

We now again define the p.d.f. of delay time as $\Omega(t)$, and its Laplace transform as $\Omega^*(s)$, which is found to be

$$\Omega^*(s) = p\bar{\alpha}_0 + \frac{p\bar{\alpha}q^*(s)\Psi_0^*(s) + (1-p)\bar{\alpha}q^*(s)}{1 - q^*(s)\Psi^*(s)} \tag{41}$$

where $q^*(s)$ is the Laplace transform of $q(t)$. From Eq. 41, we derive the mean delay as

$$\bar{t} = \frac{(1-\bar{a}_0 p)v}{\bar{\alpha}} + p\int_0^\infty t\Psi_0(t)dt + \frac{1-\bar{\alpha}_0 p}{\bar{\alpha}}\int_0^\infty t\Psi(t)dt \tag{42}$$

and the probability of no delay as

$$p_0 = p\bar{\alpha}_0 \tag{43}$$

It may be observed that the Eq. 22 for the delay caused by a sequence of cars rather than queues, is derived from Eq. 43 by setting $v = 0$, which also implies $p = 1$. Miller tested his model against some data on pedestrian crossings in Sweden, finding that his "random queue" model fitted the data slightly better than the "random car" model, and the probability of immediate crossing was predicted definitely better by the random queue model. An additional generalization of this model has been provided by Weiss (1969), who considered a mixture of cars and trucks along the main road.

Let us now consider a somewhat more complex problem. Instead of contesting against a single stream of traffic, let us assume that a driver is trying to cross a multilane highway. We first have to define a gap acceptance function α (G_1, G_2,....G_n) for crossing an n-lane highway. The first theoretical treatment of the problem was done by Tanner (1951), who assumed a negative exponential distribution of headways in each lane, and a gap acceptance function

$$\alpha(G_1, G_2, \ldots\ldots, G_n) = H(G^* - T) \tag{44}$$

where $G^* = \min_n$ ($G_1, G_2, \ldots\ldots, G_n$). Equation 44 assumes a certain "symmetry" in the treatment of all lanes. However, measurements of gap acceptance behavior for crossing two-lane highways indicate that α is not a symmetric function of the gaps (Blunden et al 1962, Wagner 1966). In fact, it is very likely that a larger gap is required by the driver for the lanes farther away from his position.

A model which allows for asymmetry in the treatment of the gaps at different lanes was proposed by Gazis et al (1967). They assumed that the gap acceptance function could be written as

$$\alpha(G_1, G_2, \ldots\ldots, G_n) = \alpha_1(G_1)\alpha_2(G_2) \ldots\ldots \alpha_n(G_n) \tag{45}$$

and that the headway distribution in each lane was a negative exponential, with the mean headway in lane j denoted by μ_j.

A general discussion of the above model is rather difficult, but useful results can be obtained if we impose some additional assumptions. Gazis et al assumed that a driver waiting at the crossing point orders the gaps according to their size, so that $G_{j_1} \le G_{j_2} \le \ldots\ldots \le G_{j_n}$. The gaps are then examined in sequence starting with the shortest one and proceeding to the longest one. Since the gap acceptance differs from lane to lane, it is possible that a short gap may be acceptable and a slightly longer one unacceptable. If G_{j_r} is the shortest unacceptable gap, then the time to the next regeneration point of the underlying renewal process is taken to be G_{j_r}, and all larger gaps are taken out of consideration.

We now let $\Omega(t)$ be the p.d.f. of the waiting time, and $\Omega^*(s)$ its Laplace transform. First of all, the probability of zero delay is the product $\bar{\alpha}_1\bar{\alpha}_2.....\bar{\alpha}_n$, where $\bar{\alpha}_r$ is the probability that the gap in lane r is acceptable, namely,

$$\bar{\alpha}_r = \frac{1}{\mu_r}\int_0^\infty \alpha_r(x)\exp\left(-\frac{x}{\mu_r}\right)dx \,. \tag{46}$$

If the delay is greater than zero, we define a regeneration point, as indicated above, as the end of the smallest unacceptable gap. Let $\rho(t)$ be the p.d.f. of the time of arrival of the first unacceptable gap, and $\rho^*(s)$ its Laplace transform. Then we derive

$$\Omega^*(s) = \bar{\alpha}_1\bar{\alpha}_2.....\bar{\alpha}_n + \rho^*(s)\Omega^*(s) \tag{47}$$

and therefore

$$\Omega^*(s) = \frac{\bar{\alpha}_1\bar{\alpha}_2.....\bar{\alpha}_n}{1-\rho^*(s)} \quad . \tag{48}$$

To complete the analysis, we need to calculate $\rho(t)$. To this end, let us suppose that the gap in lane r is unacceptable. Then, the first car to arrive in any other lane either arrives after the one in lane r, or it arrives before but the corresponding gap has been found acceptable. Therefore, we derive

$$\rho(t) = \sum_{r=1}^{\infty}\frac{1}{\mu_r}e^{-t/\mu_r}[1-\alpha(t)]\prod_{\substack{k=1\\k\neq r}}^{\infty}\left(e^{-t/\mu_k} + \frac{1}{\mu_k}\int_0^t e^{-x/\mu_k}\alpha_k(x)dx\right). \tag{49}$$

If $\alpha_r(t) = H(t-T_r)$, we can derive an expression for $\rho(t)$, and go on to find moments of the delay. Using the notation

$$\frac{1}{v_r} = \frac{1}{\mu_r} + \frac{1}{\mu_{r+1}} + ... + \frac{1}{\mu_n} \tag{50}$$

we find the following expression for the expected delay

$$\bar{t} = \exp\left(\frac{T_1}{\mu_1} + \frac{T_2}{\mu_2} + \ldots + \frac{T_n}{\mu_n}\right)\left[v_1 + (v_2 - v_1)\exp\left(-\frac{T_1}{v_1}\right)\right.$$
$$+ (v_3 - v_2)\exp\left(-\frac{T_1}{\mu_1} - \frac{T_2}{v_2}\right)$$
$$+ (v_4 - v_3)\exp\left(-\frac{T_1}{\mu_1} - \frac{T_2}{\mu_2} - \frac{T_3}{v_3}\right) + \ldots \tag{51}$$
$$\left. - (\mu_n + T_n)\exp\left(-\frac{T_1}{\mu_1} - \frac{T_2}{\mu_2} - \ldots - \frac{T_n}{\mu_n}\right)\right]$$

We can also derive an approximation to Eq. 51, which is quite satisfactory for most practical applications. The approximation consists of using the formula we derived for the delay for crossing a single lane of traffic, with $\alpha(t) = H(t - \overline{T})$, and $\varphi(t) = (1/\overline{\mu})\exp(-t/\overline{\mu})$, where the parameters $\overline{\mu}$ and $\overline{T}$ are given by

$$\frac{1}{\overline{\mu}} = \frac{1}{\mu_1} + \frac{1}{\mu_2} + \ldots + \frac{1}{\mu_n}$$
$$\overline{T} = \overline{\mu}\left(\frac{T_1}{\mu_1} + \frac{T_2}{\mu_2} + \ldots + \frac{T_n}{\mu_n}\right) \tag{52}$$

A comparison of numerical results based on Eqs. 51 and 52 shows that the approximation of Eq. 52 gives results deviating no more than 10% from those of Eq. 51. The preceding theoretical results have been further checked through simulation studies (Gazis et al 1967), which indicated that the simulation and theoretical estimates of delay were within about 5% of each other. This is not to imply that the preceding discussion provides a perfect description of the delay in merging or crossing. There are many deviations from the assumed behavior, which affect the delay. They concern both the nature of the gaps, and the behavior of drivers in accepting or rejecting them. Nevertheless, the preceding discussion gives a handle in assessing the presence of extreme cases of crossing and merging problems.

2.3.1 The "Flying Start"

An interesting variation of the preceding theme was also treated by Gazis et al. It concerns a driver who approaches the intersection at some speed and makes a decision on whether to cross or not, without stopping, (assuming, of course, that there is no stop sign at the intersection). Since the car is in motion, it is capable of crossing faster than it can from a complete stop. Therefore, the gap acceptance function at time $t = 0$ for the first gap, to be accepted with a "flying start", is different from that for following gaps, which must be accepted after a complete stop. We assume that the gap acceptance function at time $t = 0$ is

$$\alpha_0(t_1, t_2, \ldots, t_n) = \alpha_{10}(t_1) \ldots \alpha_{n0}(t_n) \tag{53}$$

where the $\alpha_{r0}(t)$ may be different from the $\alpha_r(t)$, which are used for subsequent gaps. Let $\Omega_1^*(s)$ be the Laplace transform of the p.d.f. for the delay following the first unacceptable gap. We also define the probabilities that the driver crosses with a flying start, or after the first unacceptable gap, namely,

$$\begin{aligned} \overline{\alpha}_0 &= \overline{\alpha}_{10}\overline{\alpha}_{20} \ldots \overline{\alpha}_{n0} \\ \overline{\alpha} &= \overline{\alpha}_1\overline{\alpha}_2 \ldots \overline{\alpha}_n \end{aligned} \tag{54}$$

We also define the p.d.f. for the wait to the end of the first unacceptable gap be $\rho_0(t)$, given by the expression in Eq. 49, with $\alpha_k(t)$ replaced by $\alpha_{k0}(t)$. Then, $\Omega^*(s)$ and $\Omega_1^*(s)$ satisfy the relationships

$$\begin{aligned} \Omega^*(s) &= \overline{\alpha}_0 + \rho_0^*(s)\Omega_1^*(s) \\ \Omega_1^*(s) &= \overline{\alpha} + \rho^*(s)\Omega_1^*(s) \end{aligned} \tag{55}$$

from which we obtain the solution

$$\Omega^*(s) = \overline{\alpha}_0 + \frac{\overline{\alpha}\rho_0^*(s)}{1 - \rho^*(s)} \tag{56}$$

It may be noted that Eq. 56 reduces to Eq. 48 when $\alpha_j(t) = \alpha_{j0}(t)$. If, furthermore, we assume that the gap acceptance functions are step functions, then the expected delay is derived in the same way as that given in Eq. 51. Moments of the delay time are obtained by differentiating Eq. 56 and setting $s = 0$, where the derivatives of $\rho_0^*(s)$ and $\rho^*(s)$ are also evaluated at $s = 0$. The final result is

$$\bar{t} = \bar{\alpha}_0 \bar{t}_0 + \left(1 - \bar{\alpha}_0\right)\bar{t}_1 \quad , \tag{57}$$

where $\bar{t}_0$ is given by Eq. 51 with T_j replaced by T_{j0}, and $\bar{t}_1$ is given by the same equation without any changes.

2.4 DELAYS TO PEDESTRIANS

When a pedestrian, or a group of pedestrians, arrive at a crossing point, they may be delayed by a string of cars passing by, until they find a gap large enough to allow them to cross the intersection. The behavior of pedestrians can be quite complex, depending on their age, experience, and risk aversion, but we shall use some simplifying assumptions, which have been found to describe reasonably well the queueing of pedestrians at crossing points.

One of the earliest contribution to the pedestrian queueing problem was that of Tanner (1951). Tanner assumed, basically, a single lane of traffic, although multiple lanes could be considered by using a gap acceptance function corresponding to the minimum acceptable gap. The traffic was assumed to have a negative exponential distribution, the pedestrians were assumed to arrive according to a Poisson input, and all pedestrians were assumed to have a common gap acceptance function. A more general discussion, removing some of the restrictions of the Tanner treatment, was provided by Weiss (1963, 1964), who also considered a single lane of traffic but with a general stationary headway distribution, and an unrestricted gap acceptance function. In what follows, we outline basically the Weiss treatment of pedestrian queueing.

It is rational to assume that a pedestrian arriving at an intersection, either crosses immediately, or joins a group of other waiting pedestrians and crosses with them when they perceive an acceptable gap. The Weiss solution is based on the treatment of embedded Markov chains. Transition probabilities, p_{nm}, are defined for the event that "there are m people in the queue at the regeneration point, given that n people were in the queue at the previous regeneration point". There are two possible cases in computing p_{nm}. If $n > m$, the transition $n \to m$ can only be realized in the sequence $n \to 0 \to m$; namely, the group of n crosses as a whole, and then at least m new pedestrians come to the crossing point, of which m remain in the queue. If, however, $n \leq m$, the transition $n \to m$ can either occur as $n \to 0 \to m$, or $n \to m$, since the initial group does not necessarily have to cross, but can be augmented to the size m with the addition of some new arrivals. Therefore, p_{nm} is decomposed as

$$p_{nm} = p_{nm}^{(1)} + p_{nm}^{(2)} \tag{58}$$

where $p_{nm}^{(1)}$ accounts for the indirect transition $n \to 0 \to m$, and $p_{nm}^{(2)}$ accounts for the direct transition $n \to m$. On the basis of our previous discussion, $p_{nm}^{(2)} = 0$ if $n > m$.

Let us consider a headway of t. On the basis of our assumption of a Poisson arrival of pedestrians, the probability that an arriving pedestrian will see an initial gap between x and $x+dx$ is dx/t. Therefore, the probability that a pedestrian arriving at some time within the interval $(0,t)$ will cross immediately is

$$\alpha^*(t) = \frac{1}{t}\int_0^t \alpha(x)\,dx \quad . \tag{59}$$

Let us now calculate $p_{nm}^{(1)}$, assuming that the pedestrian arrival rate is λ. If the gap is t, the pedestrian group crosses with probability $\alpha(t)$, or they do not cross, and $k+m$ pedestrians arrive with probability

$$\frac{e^{-\lambda t}(\lambda t)^{k+m}}{(k+m)!} \; .$$

Of the new arrivals, m remain in the queue and k cross the road on arrival. These assumptions imply that $p_{nm}^{(1)}$ is given by

$$
\begin{aligned}
p_{nm}^{(1)} &= \sum_{k=0}^{\infty} \int_0^{\infty} \alpha(t)\varphi(t) \frac{(\lambda t)^{k+m}}{(k+m)!} e^{-\lambda t} \binom{k+m}{m} \left[\alpha^*(t)\right]^k \left[1-\alpha^*(t)\right]^m dt \\
&= \frac{1}{m!} \sum_{k=0}^{\infty} \frac{1}{k!} \int_0^{\infty} \alpha(t)\varphi(t) \left[\lambda t \alpha^*(t)\right]^k \left[\lambda t\left(1-\alpha^*(t)\right)\right]^m dt \\
&= \frac{1}{m!} \int_0^{\infty} \alpha(t)\varphi(t) \left[T(t)\right]^m e^{-\lambda T(t)} dt
\end{aligned} \tag{60}
$$

where

$$
T(t) = \lambda t\left[1-\alpha^*(t)\right] = \lambda \int_0^t \left[1-\alpha(x)\right] dx \tag{61}
$$

Let us now consider $p_{nm}^{(2)}$, which corresponds to the case when $m \geq n$. In this case the transition $n \to m$ can take place either in the manner described by Eq. 60, or the original group does not cross, and $k+m-n$ pedestrians arrive, of whom $m-n$ remain in addition to the original group. On this basis, we obtain

$$
\begin{aligned}
p_{nm}^{(2)} &= \sum_{k=0}^{\infty} \binom{k+m-n}{m-n} \frac{1}{(k+m-n)!} \int_0^{\infty} \left[1-\alpha(t)\right]\varphi(t) e^{-\lambda t} \left(\lambda t^{k+m-n}\right) \\
&\quad \times \left[\alpha^*(t)\right]^{k+m-n} \left[1-\alpha^*(t)\right]^{m-n} dt \\
&= \int_0^{\infty} \left[1-\alpha(t)\right]\varphi(t) \frac{\left(\lambda T(t)\right)^{m-n}}{(m-n)!} e^{-\lambda T(t)} dt \; .
\end{aligned} \tag{62}
$$

If we now define the parameters

$$
\begin{aligned}
\Delta_n &= \frac{1}{n!} \int_0^{\infty} \alpha(t)\varphi(t) \left[\lambda T(t)\right]^n e^{-\lambda T(t)} dt \\
\varepsilon_n &= \frac{1}{n!} \int_0^{\infty} \left[1-\alpha(t)\right]\varphi(t) \left[\lambda T(t)\right]^n e^{-\lambda T(t)} dt
\end{aligned} \tag{63}
$$

then p_{nm} are obtainable from the transition matrix

$$P = \begin{bmatrix} \Delta_0 + \varepsilon_0 & \Delta_1 + \varepsilon_1 & \Delta_2 + \varepsilon_2 & \cdots \\ \Delta_0 & \Delta_1 + \varepsilon_0 & \Delta_2 + \varepsilon_1 & \cdots \\ \Delta_0 & \Delta_1 & \Delta_2 + \varepsilon_0 & \cdots \\ \Delta_0 & \Delta_1 & \Delta_2 & \cdots \\ \cdot & \cdot & \cdot & \\ \cdot & \cdot & \cdot & \\ \cdot & \cdot & \cdot & \end{bmatrix} \tag{64}$$

We may now proceed to derive the equilibrium properties of the underlying Markov chain defined by this transition matrix. Let $\theta = (\theta_0, \theta_1, \theta_2, \ldots.)$ be the vector of steady state probabilities, meaning that θ is the solution to the equation

$$\theta = \theta \boldsymbol{P} \quad . \tag{65}$$

Writing this equation in component form, we get

$$\theta_n = \Delta_n + \sum_{r=0}^{n} \theta_r \varepsilon_{n-r} \tag{66}$$

which can be solved by the method of generating functions. Let us define

$$\begin{aligned} \theta(z) &= \sum_{n=0}^{\infty} \theta_n z^n \\ \Delta(z) &= \sum_{n=0}^{\infty} \Delta_n z^n = \int_0^{\infty} \alpha(t)\varphi(t)e^{-\lambda T(t)(1-z)}dt \\ \varepsilon(z) &= \sum_{n=0}^{\infty} \varepsilon_n z^n = \int_0^{\infty} [1-\alpha(t)]\varphi(t)e^{-\lambda T(t)(1-z)}dt \end{aligned} \tag{67}$$

Then Eq. 66 implies that

$$\theta(z) = \frac{\Delta(z)}{1 - \varepsilon(z)} \tag{68}$$

The moments of the queue length can now be derived from these last two equations by differentiation, leading to the following expressions for the mean queue length and the associated variance

$$\begin{aligned} \bar{n} &= \lambda\mu\left(\frac{1-\bar{\alpha}_0}{\bar{\alpha}}\right) \\ \sigma^2 &= \bar{n} + \frac{\lambda^2}{\bar{\alpha}}\int_0^\infty T^2(t)\varphi(t)dt - \frac{\lambda^2}{\bar{\alpha}^2}\left[\int_0^\infty T(t)\alpha(t)\varphi(t)dt\right]^2 \end{aligned} \tag{69}$$

Assuming $\alpha(t) = H(t-T)$ and $\varphi(t) = (1/\mu)\exp(-t/\mu)$, we find that the formulas of Eq. 69 are reduced to

$$\begin{aligned} \bar{n} &= \lambda\mu\left(e^{T/\mu} - 1\right) \\ \sigma^2 &= \lambda\mu(1 + 2\lambda\mu)e^{T/\mu} - \lambda\mu\left(\frac{1 + 2\lambda\mu + 2T + \lambda T^2}{\mu}\right) \end{aligned} \tag{70}$$

which are identical to those given first by Tanner (1951).

The preceding statistics are for queue lengths at a regeneration point, but similar results can also be obtained for the statistics of queue length at a random time. In agreement with a general result by Little (1961), it is found that the mean number of pedestrians at a random time is

$$\bar{n} = \lambda\bar{t} \tag{71}$$

where $\bar{t}$ is given by the expression in Eq. 22. For a negative exponential headway distribution and a step gap distribution, we obtain the following expression for $\bar{n}$,

$$\bar{n} = \lambda\mu\left(e^{T/\mu} - 1 - \frac{T}{\mu}\right) \tag{72}$$

which was also first derived by Tanner (1951).

One may observe that there are a number of deficiencies in the underlying model of the preceding discussions. For example, the assumption that all pedestrians have the same gap acceptance function is obviously a questionable one. The question is, how much of an effect this assumption has on the results. It turns out that this particular deficiency can be circumvented quite easily. Let us suppose that the gap acceptance function depends on the gap and on a random variable v with associated p.d.f. $g(v)$, and let the rate parameter $\lambda(v)$ also depend on v. Then, Eq. 71 changes into

$$\bar{n} = \int \lambda(v)\bar{t}(v)g(v)dv \quad . \tag{73}$$

It turns out that unless very heavy traffic intensities are involved, the corrections due to a random variation in the gap acceptance function are negligible, as was shown by Weiss (1964).

Another deficiency of the presented theory is the assumption that the gap acceptance function is independent of the group size. If we introduce a set of gap acceptance functions, $\{\alpha(t)\}$, such that $\alpha_n(t)$ is the relevant function for a group n, the general problem cannot be solved in a closed form. However, if we further assume that only a finite number of $\alpha_n(t)$ are different, then a solution can be obtained. For example, if

$$\begin{aligned} \alpha_1(t) &= \beta(t) \\ \alpha_2(t) &= \alpha_3(t) = ... = \alpha(t) \end{aligned} \tag{74}$$

that is, each arriving pedestrian makes a decision to cross on the basis of $\beta(t)$, but thereafter makes decisions on the basis of $\alpha(t)$, then

$$\theta(z) = \frac{\Delta(z) + \theta_1 \gamma(z)}{1 - \varepsilon(z)} \tag{75}$$

where

$$\gamma(z) = (1 - z) \int_0^{\infty} \varphi(t)[\beta(t) - \alpha(t)] \exp[-\lambda(1 - z) T_1(t)] dt$$

$$T_1(t) = \int_0^t [1 - \beta(u)] du \tag{76}$$

$$\theta_1 = \frac{\Delta_1(1 - \varepsilon_0) + \Delta_0 \varepsilon_1}{(1 - \varepsilon_0)^2 - \gamma_1(1 - \varepsilon_0) - \gamma_0 \varepsilon_1}$$

There have been additional discussions of variations on the theme pursued above. One of them concerns crossings in which pedestrians are given preference, such as "zebra crossings", so named because of the stripes painted in the streets in the British system of this type of crossing. Smeed (1958) presented some data regarding the delay to vehicles at such crossings, involving two variables. The dependent variable was the journey time of a car over a stretch of the road covering the crossing, and the independent variable was the average number of pedestrians crossing per unit of time. A linear relationship between the two variables was found. It turns out that, if vehicular queueing effects are ignored and pedestrian flow is relatively small, this linear relationship can be found on theoretical grounds.

Let us assume that the interval over which measurements are made is large enough so that the travel time over it is a constant t_r, whether a stop is made or not, namely, the deceleration and acceleration times due to a stop are negligible. Then the total travel time is equal to $t_r + \bar{t}$, where $\bar{t}$ is the expected delay at the zebra crossing. If the arrival of pedestrians is described by a Poisson process with arrival rate λ, and the driver has a step gap acceptance function, we may use the expression given in Eq. 25 for $\bar{t}$, with $\mu = 1/\lambda$. If, furthermore, λT is small, we can expand the exponential to find the approximation $\bar{t} \approx \lambda T^2 / 2$, so that the total travel time comes out to be $t_r + \lambda T^2 / 2$, a linear function of λ, as found by Smeed (1958, 1968). Incidentally, the value of T suggested by Smeed's data lies between 5.4

and 6.5 sec. Smeed also went on to estimate the cost of pedestrian delays, in conjunction with the cost of traffic delays, with idea of leading to warrants for construction of zebra crossings.

Another interesting problem concerning pedestrian crossings is one considered by Thedeen (1969), who investigated the statistics of crossings at intersections, where there are push-button controls. The model allows for a Poisson arrival of pedestrians at a corner, where there is an indicator which can register either a red (stop) or green (go) for the pedestrian. A time t_b or greater must elapse between the end of one green period and the beginning of the next one. If a pedestrian arrives during a red period and pushes the control button, a time of at least t_a must elapse before the start of the next green phase. This time is exactly t_a if the first pedestrian arrives after a time $t_b - t_a$ (assumed positive) measured from the end of the last green phase. If he arrives at time $t \le t_b - t_a$, the next green phase starts exactly at t_b after the end of the last green phase, so that the pedestrian waits a time $t_b - t$ for the start of the next green phase. It is also assumed that a pedestrian arriving during a green period will always be able to cross. Thedeen derived three results, namely, the expected delay to a single pedestrian, the expected total pedestrian delay over a long period of time, and the expected number of pedestrians crossing during a green period. We present here the last of these three, as an example of the mathematical techniques that can be used.

Let us assume that a green period ends at $t = 0$. The first pedestrian to arrive does so either in the time interval $(0, t_b - t_a)$ or in the interval $(t_b - t_a, \infty)$. In the first instance, if the pedestrian arrives at time $t \le t_b - t_a$, the time to the next green period is $t_b - t$, and the probability that there are r arrivals in this period is $\lambda^r (t_b - t)^r \exp[-\lambda(t_b - t)]/r!$ In the second instance, if the first pedestrian arrives at $t \ge t_b - t_a$, then the time to the next green phase is t_a, so the probability of r arrivals in the period $(t, t + t_a)$ is $\lambda^r t_a^r e^{-\lambda t_a}/r!$ Hence the probability that r pedestrians arrive during a red phase is

$$p_r = \lambda \int_0^{t_b - t_a} e^{-\lambda t} \frac{\lambda^{r-1}(t_b - t)^{r-1}}{(r-1)!} e^{-\lambda(t_b - t)} dt + \lambda \int_{t_b - t_a}^{\infty} e^{-\lambda t} \frac{(\lambda t_a)^{r-1}}{(r-1)!} e^{-\lambda t_a} dt$$

$$= e^{-\lambda t_b} \left[\frac{\lambda^r (t_b^r - t_a^r)}{r!} + \frac{\lambda^{r-1} t_a^{r-1}}{(r-1)!} \right] \qquad r = 1, 2, \cdots \qquad (77)$$

Therefore, the expected number of pedestrians crossing during the combination of a red and a green phase is

$$\bar{n} = \lambda(t_b + t_g) + \exp[-\lambda(t_b - t_a)] \tag{78}$$

where t_g is the duration of the green phase. Thedeen also showed that the expected waiting time for a single pedestrian is

$$\bar{w} = \frac{t_a + \lambda(t_b^2/2)\exp[\lambda(t_b - t_a)]}{1 + \lambda(t_b + t_g)\exp[\lambda(t_b - t_a)]} \tag{79}$$

and the limiting value of the total waiting time, measured over a reasonably long total period of time, is just $\lambda t \bar{w}$. Furthermore, the long-term portion of time spent in the green phase, i.e. the fraction of the time during which cars are delayed, is

$$p_g = \frac{\lambda t_g}{\lambda(t_b + t_g) + \exp[-\lambda(t_b - t_a)]} = \frac{\bar{n}_g}{\bar{n}} \tag{80}$$

where $\bar{n}_g$ is the expected number of arrivals during the green period.

2.5 QUEUEING AND DELAYS OF VEHICLES

Let us now consider a sequence of vehicles arriving at an intersection and attempting to merge, or cross. The difference from the previous discussion is that the sequence of vehicles involves a succession of individual decisions involving gap acceptance. Unlike the case of a group of pedestrians, where all the pedestrians were assumed to cross together at an acceptable gap, this is a case where, in some instances, the number of cars crossing is smaller than the number of cars arriving, over a relatively large time period, so that the size of the queue increases indefinitely. If we assume that the arrival of cars in the feeder road is described by a Poisson process with parameter λ, then there exists a critical value λ_c, such that

- If $\lambda > \lambda_c$, the queue increases without limit, but
- If $\lambda < \lambda_c$, the queue remains finite with probability 1.

Evans, Herman, and Weiss (1964) found an expression for λ_c in the following way. The derivation requires information about the time it takes for a car to move from the second to the first position of the queue, as well as information about the gap acceptance process. Let us assume that the move-up time is Δ, which is a random variable with probability density $g(\Delta)$, and that the gap acceptance process is similar to that we assumed for the single-vehicle merging problem. We also assume that if the first driver in line merges during a gap G, the second driver in line sees a gap $G - \Delta > 0$, that is, the move-up time is always smaller than an acceptable gap. The critical rate can then be written in terms of the quantity $M(t)$, which is the expected number of cars from an infinite queue that merge in time t, if we take the limit of $M(t)$ as t goes to infinity, namely,

$$\lambda_c = \lim_{t \to \infty} \frac{M(t)}{t} . \tag{81}$$

Evans et al showed that this limit existed, and that λ_c could be calculated as the ratio of the expected cars to merge in a single gap over the expected headway. If $\alpha_n(t)$ is the probability that n cars merge during a gap of t, and $E(t)$ is the expected number of cars that merge during that gap, then by definition

$$E(t) = \sum_{n=0}^{\infty} n\alpha_n(t) . \tag{82}$$

Therefore, on the basis of its definition,

$$\lambda_c = \frac{1}{\mu} \int_0^{\infty} \varphi(t) E(t) dt \tag{83}$$

where $\varphi(t)$ is the p.d.f. of the headways, and μ is the mean headway.

The values of $\alpha_n(t)$ can now be computed recursively using the relations

$$\begin{aligned} &\alpha_0(t) = 1 - \alpha(t) \\ &\alpha_n(t) = \alpha(t) \int_0^t g(\Delta)\alpha_{n-1}(t-\Delta)d\Delta \qquad n \geq 1 \end{aligned} \tag{84}$$

where the second of Eqs. 84 is derived by assuming that the first car merges with a move-up time Δ, and is followed by a merge of exactly $n-1$ cars.

A single integral equation for $E(t)$ can be derived from Eq. 84 as

$$E(t) = \alpha(t) \int_0^t g(\Delta)[1 + E(t-\Delta)]d\Delta \ . \tag{85}$$

Equation (85) does not have a simple solution in general, but it does in the special case when we have a constant move-up time Δ_0. We then have $g(\Delta) = \delta(\Delta - \Delta_0)$, where $\delta(x)$ is the Dirac delta function, and Eq. 84 becomes

$$\begin{aligned} &E(t) = 0 && t < \Delta_0 \\ &E(t) = \alpha(t)[1 + E(t-\Delta_0)] && t > \Delta_0 \end{aligned} \tag{86}$$

The solution of Eq. 86 is

$$E(t) = \sum_{n=0}^{\infty} \alpha(t)\alpha(t-\Delta_0)\alpha(t-2\Delta_0)\cdots\alpha(t-n\Delta_0) \tag{87}$$

where $\alpha(t)$ is to be set equal to zero if its argument is negative. In particular, if $\alpha(t) = H(t-T)$, then

$$\begin{aligned} E(t) &= n+1 \quad \text{for} \quad n\Delta_0 + T \leq t \leq (n+1)\Delta_0 + T \\ &= 0 \qquad\quad \text{otherwise} \end{aligned} \tag{88}$$

This result implies that λ_c is given by

$$\lambda_c = \frac{1}{\mu}\sum_{n=0}^{\infty}(n+1)\int_{n\Delta_0+T}^{(n+1)\Delta_0+T}\varphi(x)dx = \frac{1}{\mu}\sum_{n=0}^{\infty}\phi(n\Delta_0+T) \qquad (89)$$

where $\phi(t) = \int_t^{\infty}\varphi(x)dx$. In particular, when $\phi(t) = \exp(-t/\mu)$,

$$\lambda_c = \frac{1}{\mu}\frac{\exp(-T/\mu)}{1-\exp(-\Delta_0/\mu)} \quad . \qquad (90)$$

The parameter λ_c can be found experimentally if we know the input rate, λ, on the feeder road, and the probability, p_0, that there are no cars in the queue, with the additional assumption that the queue remains finite as $t \to \infty$. To this end, the following approach was suggested by Oliver (1962). The asymptotic expected number of cars to merge in time t is $\lambda_c(1-p_0)t + o(t)$, for large values of t. By equating results in the limit $t \to \infty$, we find

$$\frac{\lambda_c}{\lambda} = \frac{1}{1-p_0} \qquad (91)$$

so that λ_c can be estimated from information available for situations with relatively low flow volumes.

The preceding discussion is the most general one for the queueing problem encountered in merging or crossing situations. The generality, of course, also entails difficulty in obtaining closed form solutions for specific situations. Several investigators have addressed such specific situations using simplified models in which some of the features of queueing are either ignored or approximated. Some of these contributions have preceded the work described above, and others followed it. Perhaps the first treatment of vehicular queueing was by Tanner (1953, 1962), who considered queues generated by two conflicting traffic streams competing for the use of a stretch of road AB wide enough for only a single vehicle. Such a situation might arise when two lanes must merge into one. It is assumed that traffic stream 1 arrives at point A as a Poisson process with a rate parameter λ_1, and stream 2 arrives at point A with a rate parameter λ_2. A vehicle of type j (j=1,2) can cross the segment AB at a constant time T_j. It is further assumed that there is a minimum time gap Δ_j between the passage of two successive vehicles of the same type through AB. Tanner drew a distinction between three cases:

(1) $T_1 > \Delta_1, \quad T_2 > \Delta_2$,

(2) $T_1 > \Delta_1, T_2 < \Delta_2$,

(3) $T_1 < \Delta_1, \quad T_2 < \Delta_2$.

The most interesting of the three, from the point of view of applications, is the first one, and it is the only one that was treated. It corresponds to the situation when either one of the queues has control of AB, and retains control as long as there are still some vehicles in that queue. The other two cases shown above allow intermittent changes of control between queues without the intervention of any external controls.

The method of solution is a rather complicated version of the technique of embedded Markov chains, and the results are rather complicated even for the special cases considered. For example, when $\Delta_1 = \Delta_2 = 0$, that is, when there is no minimum gap between successive vehicles on AB, the mean delay to a vehicle in traffic stream 1 is

$$\bar{t}_1 = \frac{e^{\lambda_2 T_1}\left\{\lambda_1 + \lambda_2 e^{(\lambda_1+\lambda_2)T_1}\right\}}{\lambda_2 e^{\lambda_2 T_1}\left\{e^{\lambda_1(T+T_{21})} - e^{\lambda_1 T_2} + 1\right\} + \lambda_1 e^{\lambda_1 T_2}\left\{e^{\lambda_2(T_1+T_2)} - e^{\lambda_2 T_1} + 1\right\}} \times \frac{e^{\lambda_2 T_2} - \lambda_2 T_2 - 1}{\lambda_2} \qquad (92)$$

with a similar result for t_2, after an interchange of the subscripts 1 and 2.

Hawkes (1965a, 1965b) treated a slightly different version of the same problem, in which the time needed to cross the segment AB is a random variable, with different distributions for the two streams. The analysis leads to a coupled pair of integro-differential equations for the joint distributions of virtual waiting times, and hence to an expression for the mean delay when the system is n equilibrium. If μ_i and v_i are the first and second moments of delay to an isolated car crossing the intersection, and $\rho_i = \lambda_i \mu_i$, then the generalization of Eq. 92 obtained by Hawkes is

$$\bar{t}_1 = \frac{\lambda_1 v_1 [(1-\rho_1) - 2\rho_2(1-\rho_1-\rho_2)] + \lambda_2 v_2 (1-\rho_1)}{2(1-\rho_1-\rho_2)(1-\rho_1-\rho_2+2\rho_1\rho_2)} \tag{93}$$

Yeo and Weesakul (1964), and Evans, Herman, and Weiss (1964) have treated variants of the Tanner model, in which the merging time is related to the critical gap appearing in a step gap acceptance function. In this case, a value of merging time is chosen from an arbitrary distribution, and this time is the critical gap in a step gap acceptance function. Consequently, both of these models take into account the possibility of a distribution of critical gaps over the population of cars. The expression obtained by Evans et al is obtained as follows. Let $g\,(t)$ be the p.d.f. of the merging time, and $g^*(s)$ its Laplace transform. Let the input of cars on the feeder road be Poisson with rate parameter λ, and let the p.d.f. for headways be $\varphi\,(G) = \exp\left(-\rho G\right)$. Let us also define a dimensionless rate parameter $\beta = \lambda\,/\,(\lambda{+}\rho)$. Then, the mean delay is obtained as

$$\bar{t} = \left(\frac{1}{\rho}\right)\frac{1}{g^*(\rho)[g^*(\rho)-\beta]}\left\{g^*(\rho) - [g^*(\rho)]^2 + \beta\rho\frac{dg^*}{dr}\right\} \tag{94}$$

where $g^*(\rho) > \beta$ for a stationary solution. In particular, if $g\,(t) = \delta(t-T)$, that is the merging time is a constant for the entire population, then

$$\bar{t} = \frac{1}{\rho}\left(\frac{1-e^{-\rho T} - \beta\rho T}{e^{-\rho T} - \beta}\right) \tag{95}$$

The mean queue length can then be computed as $n = \lambda_2 t$. One interesting conclusion in the study of Yeo and Weesakul is that for low traffic densities, the replacement of a distribution of general form for $g(t)$ by a delta function $g(t) = \delta(t - T)$ is a good approximation, but it shows poor agreement with exact results at high densities.

2.6 THE DELAY AT A TRAFFIC SIGNAL

Traffic signals, and the delay they cause to individual drivers, are of paramount importance and have received a great deal of attention. One wants to understand the delay effect of signals, and then try to minimize it by proper design of the signal cycle. The first problem is to find the delay to a single traffic stream arriving at an intersection, which handles several such streams. In order to obtain expressions for the expected delay for a single stream, it is necessary to specify the signal settings, the arrival process of the incoming stream vehicles, and the manner in which the vehicles pass through the intersection.

2.6.1 The arrival process

The starting point for many studies of delays at traffic signals has been the assumption of a simple Poisson process describing the arrival of cars. This assumption is reasonably satisfactory for very light traffic, and the absence of a nearby upstream signal which creates bunching of the cars into platoons. In heavy traffic, when the interactions between cars cannot be neglected, a generalization of the simple Poisson process is the "compound Poisson process", in which, if $N(t)$ is the number of arrivals in any interval of length t,

$$E\left[z^{N(t)}\right] = \exp\{-\Lambda t[1 - \phi(z)]\} \qquad (96)$$

If $\lambda = \Lambda\phi'(1)$, then

$$E[N(t)] = \lambda t \qquad (97)$$

$$\mathrm{var}[N(t)] = I\,\lambda t \qquad (98)$$

where I, the "index of dispersion" of the process is given by

$$I = 1 + \frac{\phi''(1)}{\lambda} \qquad (99)$$

In this case, the index of dispersion for a compound Poisson process is never less than unity, which is the value for a simple Poisson process. Some experimental studies, for example that of Miller (1964), have indicated that departures from the simple Poisson traffic are indeed in this direction, with the dispersion index often in the range of 2. Nevertheless, it would be of interest to obtain a more general and mathematically tractable process in which the dispersion index can have any value, including one less than unity. One way of doing this was suggested by Darroch[34], and may be described as follows.

In the process of binomial arrivals described in Section 2.1, let the time be regarded as consisting of discrete, contiguous intervals, each of duration h. Let the numbers of arrivals in the intervals h be independent, identically distributed random variables with a probability generating function $\varphi(z)$. This process specializes to the binomial process if

$$\varphi(z) = 1 - p + pz \quad . \tag{100}$$

However, if

$$\varphi(z) = \exp\{-\lambda h[1 - \phi(z)]\} \tag{101}$$

and if, in addition, we assume that the arrivals during a particular h interval are randomly distributed over the interval, the process reduces to the compound Poisson process described above. Since the index of dispersion for the binomial and compound Poisson process are, respectively, smaller and greater than unity, the above process generalizes the simple Poisson process in both directions of the dispersion index. In general, it may be shown that the dispersion index for Darroch's process is given by

$$I = 1 - \lambda h + \frac{\varphi''(1)}{(\lambda h)} \tag{102}$$

2.6.2 The Signal Settings and Departure Process

A period T of a fixed-cycle traffic light is divided into a "red phase", a "yellow (or amber) phase", and a green phase. Cars are not allowed to move when they view the combination of the red plus yellow phase, R , and they move during the green phase, $G = T - R$. The simplest way in which departures can be modeled, is by assuming that the cars move when they are free to do so, and the departure intervals are independent and identically distributed random variables. In the case of a single lane of traffic, it is not unreasonable to assume that the departure headways are identical, corresponding to some "saturation flow" rate. This assumption has been made in most of the literature concerning fixed-cycle lights. One may expect that the departure time of the first vehicle may be somewhat larger than that of the rest of the vehicles, but this can be accommodated by enlarging somewhat the value of R. It is also assumed that if the cars arrive during the green phase, and there is no queue, they can move on, so they incur no departure delay.

If all the vehicles do not proceed straight through the intersection, but make a right or left turn, the department times are not necessarily the same for all cars. Cars turning right may slow down somewhat, possibly unevenly, and cars turning left slow down even more, unless they are given a "green arrow" corresponding to an obstruction of the through traffic in the opposite direction, which competes for the same road space as the left-turning vehicles. In any case, it is reasonable to divide the cars into classes, and assume that, for each class, the departure times are independent, identically distributed random variables. This, of course, assumes that the classes do not interfere with each other, as is the case if there is enough storage space to accommodate slowed-down vehicles making a turn. If there is not enough space, we have a "spill-back" effect, which drastically changes the computation of delays, as discussed by Gazis (1965).

A theoretical analysis of delays due to vehicles turning contesting for roadway space with oncoming traffic was given by Newell (1959). Darroch (1964) also considered the interference of turning vehicles (and pedestrians), with the straight-on movement of vehicles. Gordon and Miller (1966) also discussed this problem, and Little (1961) evaluated the delays due to various motorists wishing to make turning movements in Poisson traffic. However, these last two references are concerned only with the delays to turning vehicles, and not the delays to oncoming traffic.

2.6.3 Evaluation of Delays

In evaluating the delay to individual cars, the following parameters must be considered. Individual cars arrive at the merging or crossing point in such a way that the instants of arrival form a renewal process. The cars are provided service in the order of their arrival, and the service times are independent, identically distributed random variables. There are three random variables of interest: the queueing time of an individual, i.e. the time spent waiting in the queue until service gets started, the number of individual cars in the system at the instants of time just after the departures of individual cars being serviced, and the busy period, i.e. the interval of time during which there is at least one car waiting. If the distributions of either the inter-arrival intervals or the service periods have an exponential tail, the distributions or the their transforms, for all three random variables, are generally obtainable, although a detailed evaluation of all these distributions may be complicated.

The traffic signal situation is complicated, compared to classical queueing theory, by the fact that during the red phase of the signal n service is possible. Another difference from the classical queueing problem is that the vehicles at the traffic signal do not necessarily depart at the order of their arrival, except in the case of a single lane of traffic. The random variable of most interest in the case of a traffic signal is the total delay to all vehicles during a cycle. The reason that we do not focus on the delay to an individual vehicle is that it is more difficult to obtain than the delay to all vehicles. This is due to the complications introduced by the red phase. The evaluation of the total delay is the most important one anyway, because it gives information about the cost of system delays, and also provides inputs for improving the settings of traffic lights. In any case, a measure of delays to individuals is provided by the average delay, obtained by dividing the total delay by the number of cars arriving during the cycle, starting with the red phase. This is assuming, of course, that the intersection is undersaturated, and all cars are services during the green phase.

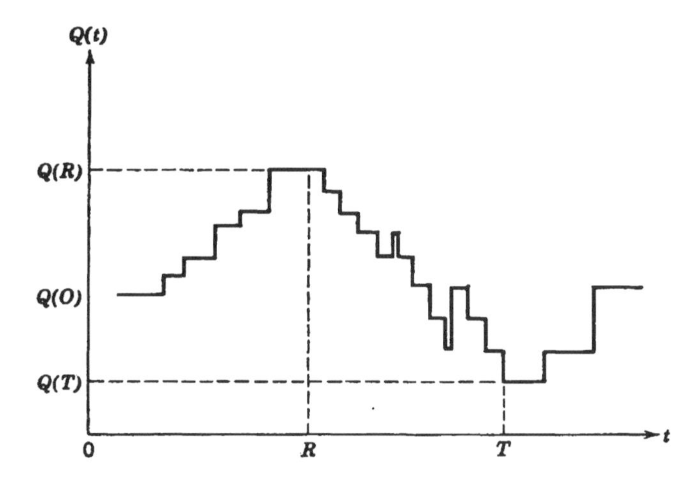

Figure 1. Queueing process at a traffic light

Figure 1 shows the queueing process leading to the creation of a queue $Q(t)$ of vehicles at the traffic light, starting at the beginning of the red phase. The delay to the vehicles during the interval $(t, t+dt)$ is $Q(t)\,\delta t + o(\delta t)$, so that the total delay to vehicles during the cycle time T is

$$W = \int_0^T Q(t)\,dt \tag{103}$$

2.6.4 Mean Delay for Poisson Input and Constant Departure Time

Let $A(t)$ be the number of vehicles which join the queue during th time interval $(0,t)$, where the origin is taken to be the start of the red phase. No departures are possible during the red phase, $(0,R)$, but departures proceed unrestricted during the green phase, (R,T). We assume that $A(t)$ is a Poisson process with $E[A(t)] = \lambda t$, and that each vehicle takes a constant time s to depart from the queue. Let

$$W = W_1 + W_2 \tag{104}$$

where

$$W_1 = \int_0^R [Q(0) + A(t)]dt, \tag{105}$$

$$W_2 = \int_R^T Q(t)dt. \tag{106}$$

The expected values in Eq. 105 are

$$E[W_1] = RE[Q(0)] + \frac{1}{2}\lambda R^2 \tag{107}$$

To find $E[W_2]$, consider first the associated variable W_2^*, defined by Eq. 106 , with the upper limit of integration set to infinity. Thus, W_2^* may be regarded as the total waiting time in a busy period for a queueing process $X(t)$ with simple Poisson arrivals, constant service times, and $X(0) = Q(R)$. It can be shown, (see, for example, McNeil 1968, and Daley and Jacobs 1969), that if $\lambda s < 1$,

$$E[W_2^*] = \frac{sE[Q(R)]}{2(1-\lambda s)^2} \tag{108}$$

Now, to avoid negligible algebraic complications, we assume that the green phase contains an integral number of departure times s , that is $T - R$ is an integer multiple of s . Since W_2 is obtained from W_2^* by neglecting that part of the queueing process after time T, we obtain, using Eq. 108,

$$E[W_2] = E[W_2^* | X(0) = Q(R)] - E[W_2^* | X(0) = Q(T)]$$

$$= \frac{sE[Q(R) - Q(T)]}{2(1-\lambda s)^2} + \frac{sE[Q^2(R) - Q^2(T)]}{2(1-\lambda s)}. \tag{109}$$

Now let us assume that the queue is in statistical equilibrium. A necessary and sufficient condition for this is that the average number of arrivals per cycle be less that the number of cars that can be services during the green phase, which is true if $\lambda T < (T - R)/s$, or

$$\lambda s < 1 - \frac{R}{T} \tag{110}$$

In this case $E\,[Q(0)] = E\,[Q(T)]$ and $E\,[Q^2(0)] = E\,[Q^2(T)]$, and $Q(R) = Q(0) + A(R)$, therefore,

$$E[Q(R) - Q(T)] = E[A(R)] = \lambda R \tag{111}$$

and

$$E[Q^2 R - Q^2(T)] = 2E[A(R)]E[Q(0)] + E[A^2(R)]$$

$$= 2\lambda R E[Q(0)] + \lambda R(\lambda R + 1)\ . \tag{112}$$

Using Equations 109, 110, and 112, we obtain

$$E[W_2] = \frac{1}{2} s(1 - \lambda s)^{-2}\{\lambda R + 2\lambda R(1 - \lambda s)E[Q(0)]\} + \lambda R(\lambda R + 1), \tag{113}$$

from which, using Eqs. 104 and 105, we obtain, for the expected total wait per cycle,

$$E[W] = \frac{2\lambda R}{2(1 - \lambda s)}\left[R + \frac{2}{\lambda} E[Q(0)] + s\left(1 + \frac{1}{1 - \lambda s}\right)\right]\ . \tag{114}$$

Dividing the right-hand side of Eq. 114 by λT, the average number of cars arriving in a cycle, we can find the mean delay to an individual vehicle. Furthermore, if the traffic is relatively light, we can ignore the term multiplied by $E[Q(0)]$, since it is unlikely that any cars remain in the queue at the end of the green phase. Otherwise, an expression for $E[Q(0)]$ must be obtained before Eq. 114 can be utilized. Before discussing this problem, we shall discuss a generalization of Eq. 114 for more general arrival and departure processes.

2.6.5 The Mean Delay for a More General Arrival Process

Let us assume that the arrival process is that postulated by Darroch (1964), while departure times remain constant. Assuming that arrivals occur at random during the h intervals, we have $E[A(t)] = \lambda t$, so Eq. 107 remains valid. To avoid some algebraic complications in seeking an expression for $E[W_2]$, we assume that s is a multiple of h, which is not an unreasonable simplification. Then, it can be shown that Eq. 108 generalizes to

$$E\left[W_2^*\right] = \frac{s(1 - \lambda s + \lambda s I)E[Q(R)]}{2(1 - \lambda s)^2} + \frac{sE\left[Q^2(r)\right]}{2(1 - \lambda s)} \tag{115}$$

where I is the index of dispersion of the arrival process $A(t)$. Consequently, Eq. 114 becomes

$$E(W) = \frac{\lambda R}{2(1 - \lambda s)}\left\{R + \frac{2}{\lambda}E[Q(0)] + s\left(1 + \frac{1}{1 - \lambda s}\right)\right\}, \tag{116}$$

which is essentially the equation obtained by Darroch. If the arrivals are binomial with $h = s$, Eq. 102 gives $I = 1 - \lambda s$, so Eq, 116 has the special form

$$E(W) = \frac{\lambda R}{2(1 - \lambda s)}\left\{R + \frac{2}{\lambda}E[Q(0)] + 2s\right\} \tag{117}$$

which is the formula obtained by Beckmann et al (1956). Actually, the formula of Beckmann et al contains an s instead of $2s$ in the right-hand side of Eq. 117, because in their model they assumed that the first departure occurred at time $s/2$ after the start of the green phase, rather than s, as assumed in our discussion. The discrepancy between the two formulas disappears if the value of $E[Q(0)]$ in the Beckmann et al discussion is increased by $\lambda s/2$, which is just the average number of arrivals in half an h interval.

Let us now consider the simplest arrival process of all, namely, the case when the vehicles in the input stream are all separated by the constant interval $1/\lambda$. This simple model was first investigated by Clayton(1941), who used it to obtain an optimal signal setting of a traffic signal which minimizes the delay at an intersection. While the process $Q(t)$ is now completely deterministic and therefore relatively trivial, it is worth considering here because it gives a measure of the minimum possible delay at an intersection. The amount by which the expected delay exceeds this minimum delay may be viewed as due to the randomness of arrivals. Clearly in this case $Q(0) \equiv 0$ and $I = 0$, so Eq. 116 becomes

$$E[W] = \frac{\lambda R}{2(1-\lambda s)}(R+s), \tag{118}$$

as obtained by Clayton. If we compare this formula to Eq. 116, we may regard the second term inside the braces as arising from a possible overflow from one cycle to the next, while the fourth term incorporates the randomness in the arrival process.

2.6.6 Mean Delay for Random Departure Times

Random departure times may be generated by differences in behavior among drivers, but also by the different time required for a right turn by some vehicles, as compared with the time for straight through movement from the same, single lane of traffic. We shall consider two important differences in this case, the effect of randomness in departure time in the computation of the expected delay, and the fact that the green phase may not coincide with the end of the departure time of a vehicle.

With respect to the effect of randomness, we assume that the departure times are independent, identically distributed random variables, with a mean value s and a coefficient of variation C. The arrival process is first assumed to be a simple Poisson process.

The expected total delay during the red phase is the same as given for constant departure times, and therefore given by Eq. 107. To find the expected total delay during the green phase, $E[W_2]$, we can use a result of Daley and Jacobs (1969), which generalizes Eq. 108 to

$$E[W_2^*]=\frac{s(1+\lambda sC^2)E[Q(R)]}{2(1-\lambda s)^2}+\frac{sE[Q^2(R)]}{2(1-\lambda s)}. \tag{119}$$

The fact that the end of the green phase does not coincide with the end of a departure time does not allow us to simply replace $Q(R)$ by $Q(T)$ in Eq. 119 and subtracting. Such a replacement yields a value which is greater than the correct value by the amount $\Delta E[Q(T)]$, where Δ is the expected difference between a departure time that starts at time T and the amount by which a departure time that ends the time period T overshoots the end of the green phase, therefore $\Delta = s - E[V]$, where V is the forward recursion time in the departure process. On the basis of renewal theory, (see, for example, Cox 1962, p.63), we find

$$E[V]=\frac{1}{2}s(1+C^2)$$
$$\Delta=\frac{1}{2}s(1-C^2) \tag{120}$$

and then, using Eq. 119, we find

$$E[W_2]=\frac{s(1+\lambda sC^2)E[Q(R)-Q(T)]}{2(1-\lambda s)^2}+\frac{sE[Q^2(R)-Q^2(T)]}{2(1-\lambda s)}+\Delta E[Q(T)]. \tag{121}$$

Equation 121, together with Eqs. 107, 111, 112, and 120, yields

$$E[W]=\frac{\lambda R}{2(1-\lambda s)}\left\{R+\frac{2}{\lambda}\left(1+\frac{(1-\lambda s)(1-C^2)}{2}E[Q(0)]\right)+s\left(1+\frac{1+\lambda sC^2}{1-\lambda s}\right)\right\} \tag{122}$$

We can observe that Eq. 122 allows us to drop the requirement of coincidence of the end of the green phase with the end of the departure time of a vehicle for all cases, including the case of constant departure times discussed earlier. Additional comments regarding the case of negative exponentially distributed departure times can be found in Weiss (1974) . A discussion of the "overflow" at the traffic signal, defined as $E\,[Q(0)]$, can also be found in Weiss.

2.6.7 Optimization of Traffic Signal Settings

The subject of optimization of traffic signal settings will be examined in some detail in the following Chapter, on Traffic Control. An extensive discussion of the optimization problem based on the queueing theory approach discussed in this chapter can be found in Weiss (1974). In particular, Weiss discusses the work of Miller (1963), who obtained analytical expressions for the optimal signal settings of an intersection.

The general idea in Miller's contribution is to compute the expected delays to competing streams of traffic, and then find the signal setting which minimizes the total delay to drivers. The problem becomes quite complex, and various approximations are introduced to make it tractable. All these approximations are justified on the basis of realistic expectations about the values of the various variables involved, concerning traffic volumes, the range of traffic signal settings, etc. One example given by Miller concerns an intersection with four competing streams, a Northbound, a Southbound, an Eastbound, and a Westbound. For such an intersection, one can obtain the rate of delay, in *vehicle hrs / hr*, for different settings of the fixed-signal. The settings are associated with the four values of π_i, which is the fraction of effective green time available to traffic stream entering from direction i. Figure 2 below shows the rate of delay obtained by Weiss for a particular example of traffic input rates. In this example, these rates are:

- Northbound 720 vehicles/hr
- Southbound 1,080 vehicles/hr
- Eastbound 180 vehicles/hr

- Westbound 360 vehicles/hr.

Clearly the North/South green setting of the traffic light must be larger than the for East/West traffic which is considerably lighter than the North/South traffic, and this setting is shown in Fig. 2 as π_{max} . The "saturation flow" , i.e. the rate of departure of vehicles from the queue, was assumed to be 1,800 vehicles/hr.

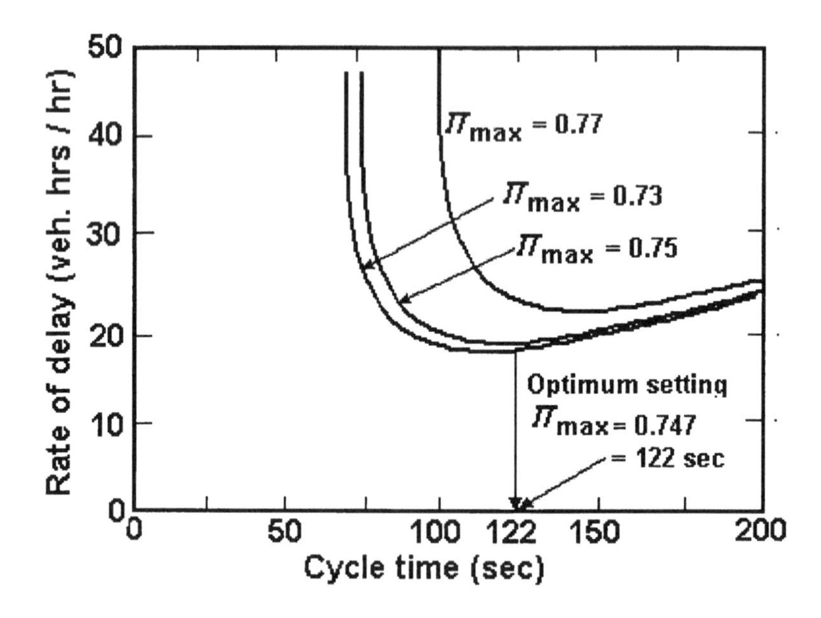

Figure 2. Delays resulting from different signal settings

It is seen from this figure that the rate of delay increases rapidly for values of the light cycle time smaller than the optimum, but less rapidly for values larger than the optimum. This means that it is wise to err on the safe side by stretching the traffic light cycle toward somewhat larger values than the computed optimum, if such a setting is realistic.

Numerous contributions to the analysis of "bulk service" queues have been made over the years, many of them aiming at improvements to the theory that would make it easier to implement. We could cite the work of Downton (1955), Gaver (1959), Miller (1959), Jaiswal (1964), Neuts (1967), Cohen (1969), Teghem et al (1969), and Chaudry (1979). A particularly useful contribution was that of Powell (1953), who proposed a method for obtaining an approximate, closed form moment formula for bulk arrival, bulk service queues. In contrast to previous work, which relied on classical transform methods, requiring the determination of complex zeros of some functions, or iterative techniques, Powell used transform results to obtain approximate, closed form formulas. The approximate formulas were obtained by solving exactly the problem for several values of the input, and then using linear regression, in which the unknown functions were the dependent variables, and some appropriate input functions were the dependent variable.

2.6.8 The Vehicle-Actuated Traffic Signal

The models we have presented thus far involve traffic lights in which the red and green phases are of fixed duration. However, there are situations, in which one direction may start getting higher input, while the cross direction still has portions of the green phase to spare. If we have the intersection properly instrumented to detect such variations in real time, we can hope to borrow some green light from the under-loaded direction and give it to the over-loaded one. For example, we can wait until the queue in the overloaded direction empties before we give the green to the cross traffic, thus utilizing the available total green time more effectively, and reducing the overall delay to users. This defines traffic actuation of signals, in which the green and red phases of both directions vary in response to actual measurements of demand. If the departure processes in the two competing directions are the same, the best policy is to wait, if possible, until the queue in the over-loaded direction empties, and then switch. The challenge, of course is to position traffic sensors properly, so that they can determine the end of the queue. This end is generally defined by measuring the headways of the arriving vehicles. It should be pointed out here that in most cases there is maximum allowable green period, intended to avoid having the vehicles in the opposing direction wait for inordinate amounts of time. (It has been said that "if a red phase is more than two minutes long, drivers assume that the light is broken down and choose to ignore it"). This limitation on the duration of the green light may negate our desire to completely exhaust the queue in the over-loaded direction, even though this may result in an increase in the expected delay to the users.

In the case of the fixed-cycle system, it was possible to determine the expected delay, and go on to optimize the signal setting, by considering each competing stream of traffic separately. This is not possible for traffic actuated signals, because the two streams must be considered together in seeking optimization. The analysis becomes even more complicated if we allow turning movements, and hence contention for green time by more than two streams. Consequently, the case of just two competing streams at an intersection of two one-way streets has received most of the attention, since it is more tractable mathematically. What is lacking from most of the studies of this problem is the application of the "California rule" of stop-and-then-go, if you have space for the "go" part.

Even in the simple case of two intersecting on-way streets, the analysis gets to be quite complicated, unless we remove the upper bound for a green phase, as was done, for example, by Tanner (1953), and by Darroch et al (1964). In both these papers, it is assumed that the arrivals are a simple Poisson process, but Darroch et al allow for variable departure times and lost times, whereas Tanner does not. Tanner's model is more realistic in dealing with the change-over of allocation of the green, at the expense of increased mathematical complexity, which does not burden the model of Darroch et al.

In the analysis that follows, we simplify matters by assuming that all the departure times have a fixed duration s, the lost times are of fixed length L, and the competing streams of traffic are a simple Poisson process with rate λ_i in the i direction. It is further assumed that the traffic signal may extend the green in either direction until the queue is exhausted, which is determined by the detection of a headway of at least β_i in the arrivals near the end of the queue. This is a special case of the Darroch et al model, but their generalization to independent and identically distributed departure times, possibly with different distributions for the two competing streams, is straight-forward. The optimization problem consists of finding the values of β_i which minimize the total rate of delay at the intersection. Since there is no "overflow" at the end of the green phase, with headways exceeding β_i, the expected total waiting time per cycle is given by Eq. 116, without the term involving $E\,[Q\,(0)]$. The red period is now a random variable, and so we have, in equilibrium,

$$E[W_i] = \frac{\lambda_i}{2(1-\lambda_i s)}\left\{E\left[R_i^2\right] + sE[R_i]\left(1 + \frac{1}{1-\lambda_i s}\right)\right\} . \tag{123}$$

In order to evaluate the first two moments of R_i, (Fig. 3), we consider two successive light cycles and obtain

$$\begin{aligned} R_1^{(j+1)} &= 2L + Y_2^{(j)} + Z_2^{(j)} - \beta_1 \ , \\ R_2^{(j)} &= 2L + Y_1^{(j)} + Z_1^{(j)} - \beta_2 \ , \end{aligned} \tag{124}$$

where Y_i is the time it takes to discharge the queue i, Z_i is additional time that elapses before the beginning of a headway of at least β_i, and the superscript j denotes the cycle number.

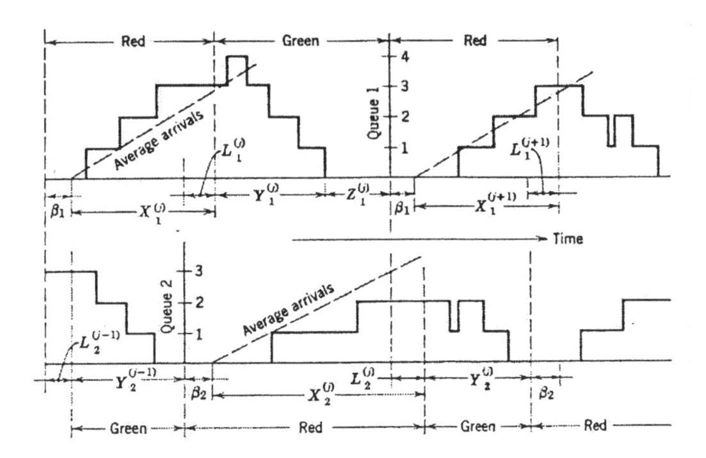

Figure 3. Traffic queues at vehicle-actuated traffic signals

Suppose that there are Q_i vehicles queued in the direction i at the beginning of its green phase. Then Y_i is the convolution of $Q_i^{(j)}$ busy periods in a queue with simple Poisson arrivals with rate λ_i and constant service times of duration s. We can now use known results from queueing theory (see, for example, Cox and Smith 1961, p. 55) and obtain

$$E[Y_i|Q_i] = \frac{Q_i s}{1-\lambda_i s},$$
$$E[Y_i^2|Q_i] = E^2[Y_i|Q_i] + \frac{Q_i \lambda_i s^3}{(1-\lambda_i s)^3}. \tag{125}$$

Now, $Q_i^{(j)}$ is the number of arrivals in the red period $R_i^{(j)}$, and therefore it is Poisson distributed with mean $\lambda_i R_i^{(j)}$. Therefore, taking expectations in Eq. 125, we obtain

$$E[Y_i] = \frac{\lambda_i s E[R_i]}{1-\lambda_i s} \; ,$$
$$E[Y_i^2] = \lambda_i s^2 \frac{\{E[R_i] + \lambda_i (1-\lambda_i s) E[R_i^2]\}}{(1-\lambda_i s)^3} \; . \tag{126}$$

Now, taking expected values in Eq. 124 and using Eq. 126, we obtain

$$E[R_1] = 2L - \beta_1 + E[Z_2] + \frac{\lambda_2 s E[R_2]}{1-\lambda_2 s} \; ,$$
$$E[R_2] = 2L - \beta_2 + E[Z_1] + \frac{\lambda_1 s E[R_1]}{1-\lambda_1 s} \; . \tag{127}$$

Solving Eq. 127 for the two $E[R_i]$, we obtain

$$E[R_1] = \frac{(1-\lambda_1 s)\{2L - \beta_1 + \lambda_2 s(\beta_1 - \beta_2) + \lambda_2 s E[Z_1] + (1-\lambda_2 s) E[Z_2]\}}{1-(\lambda_1 + \lambda_2) s} \; , \tag{128}$$

and a similar expression for $E[R_2]$ with the subscripts 1 and 2 interchanged in Eq. 128, for reasons of symmetry.

Next, we obtain the average cycle time. From Fig. 3, we see that

$$E[T] = E[R_1] + E[R_2] - 2L + \beta_1 + \beta_2 \; . \tag{129}$$

From Eqs. 128 and 129, we obtain

$$E[T] = \frac{2L + (1-\lambda_1 s) E[Z_1] + (1-\lambda_2 s) E[Z_2] - s(\beta_1 \lambda_1 + \beta_2 \lambda_2)}{1 - s(\lambda_1 + \lambda_2)} \; . \tag{130}$$

The values of $E[R_i^2]$ can be obtained by squaring both sides of Eq. 124 and taking expectations. Since Y_i and Z_i are independent, we obtain

$$E[R_1^2] = (2L - \beta_1)^2 + 2(2L - \beta_1)\{E[Y_2] + E[Z_2]\} \\ + E[Y_2^2] + 2E[Y_2]E[Z_2] + E[Z_2^2] \tag{131}$$

and a similar expression for $E[R_2^2]$ with the subscripts 1 and 2 interchanged. Explicit expressions for all the expected values of variables can now be obtained using Eqs. 126 and 128, but they are too complicated to show here.

To obtain the total rate of delay per unit of time, i.e. $\{E[W_1] + E[W_2]\}/E[T]$, we have to determine the first two moments of Z_i, which is just the waiting time for stream i until the sensing of the first gap of at least β_i duration , and therefore, using Eqs. 25 we obtain

$$E[Z + \beta_i] = \frac{\exp(\lambda_i\beta_i) - 1}{\lambda_i} \\ E[(Z_i + \beta_i)^2] = 2\exp(\lambda_i\beta_i)\frac{\exp(\lambda_i\beta_i) - 1 - \lambda_i\beta_i}{\lambda_i^2} \quad . \tag{132}$$

It can now be shown that the condition for equilibrium is, as expected,

$$(\lambda_1 + \lambda_2)s < 1 \quad . \tag{133}$$

Next, we can determine the values of β_{ii} which minimize the total delay per unit of time. In general, graphical or numerical methods are required for this task, except in special cases. Interestingly enough, the derivation of the rate of delay is simplified considerably if the intersection is operating close to saturation, that is, if $(\lambda_1 + \lambda_2)s$ is very close to unity. In such a case, the right-hand side of Eq. 128 is very large, since its denominator is very small. This results in $E[Y_i]$ and $E[Y_i^2]$ being the dominant terms in the right-hand side of Eq. 131. Neglecting the other terms of the right-hand side and using Eq. 126, we obtain

$$E\left[R_1^2\right] \approx \frac{\lambda_2^2 s^2}{(1-\lambda_2 s)^2} E\left[R_2^2\right] + \left\{ \frac{2(2L - R_1) + E[Z_2]}{1-\lambda_2 s} + \frac{\lambda_2 s^2}{(1-\lambda_2 s)^3} \right\} E[R_2] , \tag{134}$$

and a similar expression for $E\left[R_2^2\right]$ with the subscripts 1 and 2 interchanged. In addition, if $(\lambda_1 + \lambda_2)s$ is close to unity, Eqs. 128 and 130 yield

$$\begin{aligned} E[R_1] &\approx \lambda_2 s E[T] , \\ E[R_2] &\approx \lambda_1 s E[T] . \end{aligned} \tag{135}$$

We now solve Eq. 134 and its counterpart for $E\left[R_2^2\right]$ as simultaneous equations, and using Eq. 135 we obtain

$$\lambda_i^2 E\left[R_i^2\right] \approx \frac{\lambda_1 \lambda_2 s^2 E[T]}{2[1-(\lambda_1+\lambda_2)s]} \{(\lambda_1+\lambda_2)(1+4\lambda_1\lambda_2 Ls)$$

$$+ 2\lambda_1\lambda_2 s(\lambda_1 E[Z_2] + \lambda_2 E[Z_1] - \lambda_1\beta_1 - \lambda_2\beta_2) \}$$

Using Eq. 130, we can further reduce this expression to

$$\lambda_i^2 E\left[R_i^2\right] \approx \lambda_1 \lambda_2 E[T] \left\{ E[T] + \frac{s - 4L\left(1-\lambda_1\lambda_2 s^2\right)}{2[1-(\lambda_1+\lambda_2)s]} \right\} . \tag{136}$$

From Eqs. 123 and 136 we now obtain a good approximation for the rate of delay, as $(\lambda_1 + \lambda_2)s$ tends to unity, given by

$$\frac{E[W_1]+E[W_2]}{E[T]} \approx \frac{E[T]}{s} + \frac{1-4(1-\lambda_1\lambda_2 s^2)L/s}{2[1-(\lambda_1+\lambda_2)s]} \quad . \tag{137}$$

The right-hand side of Eq. 136 only involves the values of β_i through the mean cycle time $E[T]$. The values of β_i for which $E[T]$ is minimum are obtainable from $\partial E[T]/\partial \beta_i = 0$. Entering Eqs. 132 into Eq. 130 and differentiating, we obtain the optimal values

$$\beta_i = \frac{1}{\lambda_i}\log\left(\frac{1}{1-\lambda_i s}\right) \quad . \tag{138}$$

subject to the additional requirement that $\beta_i \leq 2L$, Eq. 138 gives the values of β_i which minimize the total rate of delay, at high traffic volumes. It should be pointed, however, that the usefulness of the result may be limited by constraints on the maximum value of the red phase, in which case one may seek settings which are as close to the optimum obtained above as possible.

A recent review of queueing problems in transportation has been provided by Hall (1999). In this review, Hall discusses delays not only in roadway traffic situations, but also in railroad train movements, covering additional aspects of queueing in transportation, with an emphasis on "bulk service models", which arise naturally in traffic because of the consolidation function performed by the vehicles. A particularly interesting contribution to the optimization of the management of bulk arrival, bulk service queues has been given by Powell (1986). Powell bypasses the honerous task of evaluating complex zeros of functions derived from transform methods used in queueing studies by seeking suitable approximations to the equations involved. He obtains closed form expressions for the mean and variance of the length of the queue at departure instants, as well as the moments of the length of the queue in continuous time, and of the waiting time distribution.

References

Adams, W. F. (1936), "Road Traffic Considered as a Random Series", *J. Inst. Civil Engrs.*, **4,** 121-130.

Beckmann, M. J., C. B. McGuire, C. B. and C. B. Winsten, C. B. (1956), *Studies in the Economics of Transportation* , New Haven: Yale University Press.

Blunden, W. R., Clissold, C. M. and Fisher, R. B. (1962), "Distribution of Acceptance Gaps for Crossing and Turning Maneuvers", *Proc. Aust. Road Res. Board*, **1**, 188-205.

Breiman, L. (1963), "The Poisson Tendency in Traffic Distribution", *Ann. Math. Statist.*, **34**, 308-311.

Chaudry, M. L. (1979), " The Queueing System $M^* / G / 1$ and its Ramifications", *Naval Res. Logist. Quart.*, **26**, 667-674.

Clayton, A. J. H. (1941), "Road Traffic Calculations", *J. Inst. Civil Engrs.*, **16**, No. 7, 247-284 (1941); **16**, No. 8, 588-594.

Cohen, J. W. (1969), *The Single Server Queue*, London: North Holland.

Cox, D. R. and Smith,, W. L. (1961), *Queues*, London: Methuen.

Cox, D. R. (1962), *Renewal Theory*, London, Methuen.

Daley, D. J. and Jacobs, D. R. (1969), "The Total Waiting Time in a Busy period of a Stable Single-Server Queue, II), *J. Appl. Prob.*, **6**, 573-583.

Darroch, J. N. (1964), "On the Traffic-Light Queue", *Ann. Math. Statist.*, **35**, 380-388.

Darroch, J. N., Newell, G. F. and Morris, R. W. J. (1964), "Queues for a Vehicle-Actuated Traffic Light", *Operations Res.*, **12**, 882-895.

Downton, F. (1955), "Waiting Time in Bulk Service Queues", *J. Roy. Stat. Soc.*, *B* **17**, 256-261.

Drew, D. R., LaMotte, L. R., Buhr, J. H. and Wattleworth, J. A. (1967), "Gap Acceptance in the Freeway Merging Process", Report 430-2, Texas Transportation Institute.

Drew, D. R., Buhr, J. H. and Whitson, R. H. (1967), "The Determination of Merging Capacity and its Applications to Freeway Design and Control", Report 430-4, Texas Transportation Institute.

Evans, D. H., Herman, R. and Weiss, G. H. (1964), "The Highway Merging and Queueing Problem", *Operations Res.*, **12,** 832-857.

Gaver, D. P. (1959), "Imbedded Markov Chain Analysis of a Waiting Line Process in Continuous Time", *Ann. Math. Stat.*, **30**, 698-720.

Gazis, D. C. (1965), "Spill-back from an Exit Ramp of an Expressway", *Highway Research Record*, **89**, 39-46.

Gazis, D. C., Newell, G. F., Warren, P. and Weiss, G. H. (1967), "The Delay Problem for Crossing an *n* Lane Highway", *Proc. 3rd Intern. Symp. on Traffic Flow*, New York, Elsevier, pp. 267-279.

Gordon, J. D. and Miller, A. J. (1966), "Right-Turn Movements at Signalized Intersections", *Proc. Third Conf. Aus. Road Res. Board*, **1**, 446- 459.

Grace, M. G., Morris, R. W. J. and Pak-Poy, P. G. (1964), "Some Aspects of Intersection Capacity and Traffic Signal Control by Computer Simulation", *Proc. Second Conf. Aus. Road Res. Board*, **1**, 274-304.

Hall, R. W. (1999), "Transportation Queueing", in *Handbook of Transportation Science*, edited by R. W. Hall, Boston: Kluwer Academic Publishers.

Hawkes, A. G. (1965a), "Queueing at Traffic Intersections", *Proc. 2nd Intern. Symp. on the Theory of Traffic Flow*, edited by J. Almond, Paris: Office of Economic Cooperation and Development, pp. 190-199.

Hawkes, A. G. (1965b), "Queueing for Gaps in Traffic", *Biometrica*, **52**, 79-85.

Herman, R. and Weiss, G. H. (1961), "Comments on the Highway Crossing Problem", *Operations Res.*, **9**, 828-840.

Jaiswal, N. K. (1964), "A Bulk Service Queueing Problem with Variable Capacity", *J. Roy. Stat. Soc.*, *B* **26**, 143-148.

Little, J. D. C. (1961), "Approximate Expected Delays for Several Maneuvers by a Driver in Poisson Traffic", *Operations Res.*, **9**, 39-52.

McNeil, D. R. (1968), "A solution to the Fixed Cycle Traffic Light Problem for Compound Poisson Arrivals", *J. Appl. Prob*, **5**, 624-635.

Miller, A. J. (1961), "A Queueing Model for Road Traffic Flow", *Proc. Roy. Statis. Soc.*, **B-23**, 64-90.

Miller, A. J. (1963), "Setting for Fixed-Cycle Traffic Signals", *Operations Res.*, **14**, 373-386.

Miller, A. J. (1964), "Settings for Fixed-Cycle Traffic Signals", *Proc. 2nd Conf. Aus. Road Res. Board*, **1**, 342-365.

Miller, R. G. (1959), "A Contribution to the Theory of Bulk Queues", *J. Roy. Stat. Soc.*, *B* **21**, 320-337.

Neuts, M. F. (1967), "A General Class of Bulk Queues with Poisson Input", *Ann. Math Stat.*, **38**, 759- 770.

Newell, G. F. (1959), "The Effects of Left Turns on the Capacity of Road Intersections", *Quart. Appl. Math.*, **17**, 67-76.

Oliver, R. M. (1962), "Distribution of Gaps and Blocks in a Traffic Stream", *Operations Res.*, **10**, 197-217.

Powell, W. B. (1986), "Approximate Closed form Moment Formulas for Bulk Arrival, Bulk Service Queues", *Transp. Sci.*, **20**, 13-23.

Schuhl, A. (1955), "The Probability Theory Applied to Distributions of Vehicles on Two-Lane Highways", *Poisson and Traffic*, New Haven: Eno Foundation, pp. 59-72.

Smeed, R. J. (1958), "Theoretical Studies and Operational Research on Traffic and Traffic Cogestion", *Bull. Intrn. Statist. Inst.*, **36**, 347-375.

Smeed, R. J. (1968), "Aspects of Pedestrian Safety", *J. Transport Econ. Policy*, **2,** No. 3, 1-25.

Tanner, J. C. (1951), "The Delay to Pedestrians Crossing a Road", *Biometrika*, **38**, 383-392.

Tanner, J. C. (1953), "A Problem of Interference Between Two Queues", *Biometrica* , **40**, 58-69.

Tanner, J. C. (1962), "A Theoretical Analysis of Delays at an Uncontrolled Intersection", *Biometrica*, **49**, 163-170.

Teghem, J. Loris-Teghem, J. and Lambotte, J. P. (1969), *Modeles d' attente $M/G/1$ et $GI/M/1$ a Arrivees et Services en Groups,* Lecture notes in Operations Research and Mathematical Economics, 8, New York: Springer-Verlag.

Thedeen, T. (1964), "A Note on the Poisson Tendency in Traffic", *Ann. Math. Statist.*, **35**, 1823-1824.

Thedeen, T. (1969), "Delays at Pedestrian Crossings of the Push-Button Type", *Proc. 4th Symp. on Theory of Traffic Flow*, Bonn: Bundesminister fur Verker, pp. 127-130.

Wagner, F. A. (1966), "An Evaluation of Fundamental Driver Decisions and Reactions at an Intersection", *Highway Res. Rec.*, **118**, 68-84.

Weiss, G. H. and Herman, R. (1962), "Statistical Properties of Low Density Traffic", *Quart. Appl. Math.*, **20**, 121-130.

Weiss, G. H. and Maradudin, A. A. (1962), "Some Problems in Traffic Flow", *Operations Res.*, **10**, 74-104.

Weiss, G. H. (1963), " An Analysis of Pedestrian Queueing", *J. Res. Natl. Bur. Stds.*, **67B**, 229-243.

Weiss, G. H. (1964), "Effects of a Distribution of Gap Acceptance Functions on Pedestrian Queues", *J. Res. Natl. Bur. Stds.*, **68B**, 13-15.

Weiss, G. H. (1966), "The Intersection Delay Problem with Correlated Gap Acceptance", *Operations Res.*, **14**, 614-619.

Weiss, G. H. (1967), "The Intersection Delay Problem with Gap Acceptance Function Depending on Space and Time", *Transport. Res.*, **1**, 367-371.

Weiss, G. H. (1969), "The Intersection Delay Problem with Mixed Cars and Trucks", *Transport. Res.*, **3**, 195-199.

Weiss, W. H. (1974), "Delay Problems for Isolated Intersections", Chapter 2 in *Traffic Science*, edited by D. C. Gazis, New York: John Wiley and Sons.

Yeo, G. F. and Weesakul, B. (1964), "Delays to Road Traffic at an Intersection", *J. Appl. Prob.*, **2**, 297-310.

Additional References

Ashworth, R. (1970), "The Analysis and Interpretation of Gap Acceptance Data", *Transp. Sci.*, **4**, 270-280.

Blumenfeld, D. E. and Weiss, G. H. (1970a), "On the Robustness of Certain Assumptions in the Merging Delay Problem", *Transp. Res.*, **4**, 125-139.

Blumenfeld, D. E. and Weiss, G. H. (1970b), "On Queue Splitting to Reduce Waiting Time", *Transp. Res.*, **4**, 141-144.

Blumenfeld, D. E. and Weiss, G. H. (1971), "Merging from an Acceleration Lane", *Transp. Sci.*, **5**, 161-168.

Buckley, D. J. (1968), "A Semi-Poisson Model of Traffic Flow", *Transp. Sci.*, **2**, 107-133.

Burns, L. D., Hall, R. W., Blumenfeld, D. E. and Daganzo, C. F. (1965), "Distribution Strategies that Minimize Transportation and Inventory Cost", *Operations Res.*, **33**, 469-490.

Cheng, T. E. C. and Allan, S. (1992), "A Review of Stochastic Modeling of Delay and Capacity at Unsignalized Intersections", *European J. of Operations Res.*, **68**, 247-259.

Dunne, M. C. (1967), "Traffic Delay at a Signalized Intersection with Binomial Arrivals", *Transp. Sci.*, **1**, 24-31.

Edie, L. C. (1956), "Traffic Delays at Toll Booths", *Operations Res.*, **4**, 107-138.

Gibbs, W. L. (1968), "Driver Gap Acceptance at Intersections", *J. Appl. Psych.*, **52**, 200-211.

Grace, M. M. and Potts, R. B. (1964), "A Theoty of Diffusion of Traffic Platoons", *Operations Res.*, **12**, 255-275.

Grace, M. G., Morris, R. W. J. and Pak-Poy, P. G. (1964), "Some Aspects of Intersection Capacity and Traffic Signal Control by Computer Simulation", *Proc. Second Conf. Aus. Road Res. Board*, **1**, 274-304.

Haight, F. A., Bisbee, E. F. and Wojcik, C. (1962), "Some Mathematical Aspects of the Problem of Merging", *Highway Res. Board Bull.*, **356**.

Hall, R. W. (1991), *Queueing Methods for Services and Manufacturing*, Englewood Cliffs, NJ: Prentice Hall.

Lindley, D. V. (1952), "The Theory of Queues with a Single Server", *Proc. Comb. Phil. Soc.*, **48**, 277-289.

May, A. D. and H. E. Keller, H. E. (1967), "A Deterministic Queueing Model", *Transp. Res.*, **1**, 117-128.

Miller, A. J. (1972), "Nine Estimators of Gap-Acceptance Parameters", *Proc. 5th Intern. Symp. on the Theory of Traffic Flow ans Transp.*, G. F. Newell, Ed., New York: Elsevier, pp. 215-235.

Newell, G. F. (1965), "Approximation Methods for Queues with Application to the Fixed-Cycle Traffic Light", *SIAM Rev.*, **7**, 223-240.

Newell, G. F. (1982), *Applications of Queueing Theory*, London: Chapman and Hall.

Oliver, R. M. and Bisbee, E. F. (1962), "Queueing for Gaps in High Flow Traffic Streams", *Operations Res.*, 105-114.

Raff, M. S. (1951), "The Distribution of Blocks in an Uncongested Stream of Automobile Traffic", *J. Amer. Statist. Assoc.*, **46**, 114-123.

Solberg, P. and Oppenlander, J. D. (1966), "Lag and Gap Acceptances at a Stop-Controlled Intersection", *Highway Res. Rec.*, **118**, 68-84.

Wardrop, J. G. (1968), "Journey peed and Flow in Central Urban Areas", *Traffic Engr. Cont.*, 528-532.

Weiss, G. H. (1964), "Effects of a Distribution of Gap Acceptance Functions on Pedestrian Queues", *J. Res. Nat. Bur. Stds.*, **68B**, 13-15.

Chapter 3

Traffic Control

Traffic problems, and measures of traffic control intended to reduce these problems, predate the automobile. In "*Famous First Facts*" by Kane (1964), we find information related to traffic problems encountered by horse driven carriages in 1791. A New York City ordinance at the time proclaimed: "Ladies and Gentlemen will order their Coachmen to take up and set down with their Horse Heads to the East River, to avoid confusion". Thus, the first one-way street was implemented. For many year after that, traffic problems continued to intensify, partially reinforced by the mixture of traffic the road facilities were expected to serve. The automobile firmly established the existence of traffic problems, and the steady deterioration of traffic networks, insofar as driver comfort is concerned. There have been brief periods when, in some locations, improvements in traffic systems were catching up with traffic, but over about one century of the existence of the automobile, the traffic situation has gone from bad to worse.

Traffic control came about mostly for reasons of safety, to avoid coincidence of two cars in space at the same time. The earliest devices of traffic control were hand signals of traffic policemen, sometimes mistaken by early motorists as friendly greetings. The first traffic light was installed in Cleveland in 1914, (see Marsh, 1927), although Detroit has claimed being the first, with a "first traffic light" shown at the Ford Museum in Dearborn, Michigan. Traffic lights proliferated, and the evolution of signals operating under advanced design concepts is shown in Fig. 1. The evolution was intended to accomplish something beyond safety, namely, to reduce unnecessary stops by individual cars and consequently reduce delays and the overall travel time required for a trip.

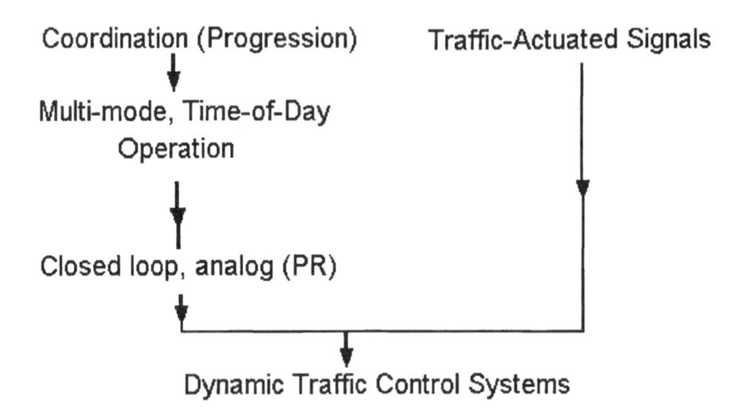

Figure 1. Evolution of traffic signal systems

Coordination of a string of traffic signals was intended to allow cars to move without stops through a sequence of intersections, preferably in both directions. The first such "progression" system was installed in Salt Lake City, in 1918, (see Marsh, 1927). Later on, the concept of the "multi-dial" system came along. The idea was to adjust progression settings for different traffic conditions, given the habits of the driving population. The multi-dial system generally provided three settings, one for morning rush hour, one for evening rush hour, and one for average conditions. Later, attempts were made to make the transition from one synchronization scheme to another responsive to actual measurements of traffic. The best known example of such a system was the "PR System" installed in Baltimore in the early 1950s.

On the other side of the picture, "traffic actuated signals" were introduced, whose settings were adjusted to respond to actual traffic measurements taken by sensors. Some of these traffic- actuated signals detect traffic only on a secondary street intersecting a main street, and keep the main street traffic moving until some traffic appears on the secondary street. Others measure traffic in both competing directions and allocate "green phases" to these competing streams appropriate for their respective volumes. All of these signals were designed for isolated intersections.

Computerized traffic control was introduced in the mid-1960s, with leading projects in Toronto, Canada, and San Jose, California, (see Gazis and

Bermant, 1966, and *San Jose Project,* 1966). In these computerized systems, progression choice was based on actual measurements of traffic, and it was coordinated with traffic-actuated control of "critical intersections". Finally, in recent years, the concept of "Intelligent Transportation Systems (ITS" has been introduced. An ITS system is supposed to optimize the design of traffic lights in a network, and also inform the drivers of events that influence the performance of the traffic system, such as accidents and other "incidents".

In this chapter, we start with a discussion of the objectives of traffic control. We then discuss the elements of traffic control, including the optimization of the operation of a single traffic light, or a group of traffic lights, optimal synchronization schemes, and finally optimal operation of control devices in a network. Particular attention is paid to the issue of congestion, which causes networks to function as "store-and-forward" systems, in which queueing is virtually inevitable, and the throughput of the network depends on the throughput of critical points defined as "Throughput Limiting Points (TLPs)".

3.1 OBJECTIVES OF TRAFFIC CONTROL

There are two main objectives in traffic control. One is *Safety*, and the other one is *Comfort and Convenience of Drivers.* The safety objective is obvious, and it consists of separating streams of traffic to avoid collisions, and instructing drivers on points of danger, such as sharp curves. The objective of maximizing the comfort and convenience to drivers comprises means for allowing drivers to complete their journey in as short time as possible, and means for easing the driving task, for example by reducing the number of unnecessary stops. Viewed as a control theory problem, traffic control entails the goal of improving an *objective function* , subject to constraints of the traffic system. Optimization is to be achieved by selecting any of a number of controllable parameters of the system. Such parameters include the *split* , i.e. division of the green phase among competing streams, the *offset* of a sequence of lights, i.e. the time difference between the start of the green phase of these lights, etc.

The objective function of the optimization problem can be any combination of the following measures of performance of a network:

- Average delay per vehicle,
- Maximum individual delay,
- Percentage of cars that are stopped as they traverse the system,

- Average number of stops for an average trip,
- Throughput of intersections or the system at large, and
- Travel time through the system (average or maximum).

The average delay, or equivalently the "aggregate delay" to all users, is the most commonly used objective function, and it is related to the last item listed above. The other criteria of performance listed above are used either as constraints in the optimization process, or as objective functions in special situations. For example, minimizing the average number of stops is the objective of the design of progression, although it is not explicitly stated as an objective function.

We shall discuss the following traffic systems, which have been investigated over the years:

- An isolated intersection,
- A system of intersections amenable to signal coordination, or coordinated control for optimizing the objective function,
- Critical traffic links, such as tunnels, bridges, and special freeway sections, and
- An urban street network.

Since the onset of computerization of traffic control, some of the above systems have been considered in combination. For example, traffic signals in a network have been designed to accomplish two types of control, *macrocontrol*, (or *major loop control*), and *microcontrol*, (or *minor loop control*). The idea is to devise an overall strategy of coordination of traffic lights in a network, which makes the largest possible contribution to the optimization process, e.g. toward minimizing delays to drivers. Such a global strategy constitutes the macrcontrol. Over and above that, microcontrol is instituted, comprising the optimal management of "critical points" in the network, generally in response to actual real-time measurements of traffic in the vicinity of these points.

The pursuit of ITS has, in principle, integrated all the above objectives, and added a few others, as will be discussed later. One of the distinguishing points of ITS is the desire to communicate with drivers, in order to alert them as to the current traffic conditions. Such communication is expected to influence the route choices of drivers, thus affecting the inputs to the optimization problem for the network.

3.2 SINGLE, ISOLATED INTERSECTION

Let us consider an isolated intersection serving two conflicting streams perpendicular to each other. The intersection may actually be serving four streams, i (= 1,2), two in the north/South and two in the East/West directions, but it is assumed that two conflicting streams are dominant at a particular time of the day, a normal occurrence in urban traffic. We define the following quantities

- q_i = Traffic flow volume along the i direction
- s_i = Saturation flow along the i direction
- g_i = Duration of green phase favoring stream i.
- c = Common cycle of all intersection along an artery

The saturation flow is defined as the flow achieved when a platoon of vehicles crosses the intersection, starting from a complete stop. It is tacitly assumed that individual vehicles depart with the same headway, but a slight variation from vehicle to vehicle would not alter the results substantially. We define, in addition, the traffic cycle as c , and the total lost time during acceleration and clearing of the intersection, (mostly equal to the yellow phase), as L .

We can now use the following reasoning: During the green phase g_i , the throughput in the i direction is $s_i g_i$. It is reasonable to set the green phases in a way that provides equal saturation levels, $q_i / s_i g_i$, in both directions, where of course it is assumed that these saturation levels, and their sum, are smaller than unity. This yields to the empirical formula[5] for the split of the effective green

$$\frac{q_1}{s_1 g_1} = \frac{q_2}{s_2 g_2} \tag{1}$$

where the green phases in the two directions satisfy the relationship

$$g_1 + g_2 + L = c \ . \tag{2}$$

The above relationships have been the bible of traffic engineers for ever. Webster (1958) investigated the optimum setting of traffic lights under realistic assumptions about arrival patterns and queue discharge during the green phase, and obtained the optimum traffic cycle, and "split" of the green between the two directions. Webster found that the split defined by Eqs. 1,2 was pretty close to the one obtained by his optimization process, in most practical situations.

As was mentioned already, the saturation levels in the two directions should add up to less than unity. In fact, we must have

$$\frac{q_1}{s_1}+\frac{q_2}{s_2}<1-\frac{L}{c} \tag{3}$$

The condition of Eq. 3 guarantees that the intersection is "undersaturated". The case of the "oversaturated" intersection is discussed later in this chapter.

The delay to drivers at a single intersection can be further reduced if one takes into account the variation of inputs. This can be achieved in practice with proper instrumentation measuring the traffic inputs, leading to the "traffic actuated" operation of the traffic light. Dunn and Potts (1964) have developed an adaptive control algorithm based on such real-time measurements of traffic input.

3.3 SYNCHRONIZATION SCHEMES FOR ARTERIAL TRAFFIC

The objective of properly synchronizing a string of traffic signals along an artery is to allow as many drivers as possible to go through the sequence of traffic lights without stopping, preferably in both directions. This provides "comfort and convenience" to drivers, the comfort reflected by the reduced number of stops, and the convenience by reduced travel times. It is assumed, at this point, that the system in "undersaturated, that, is, every intersection has adequate aggregate green time to serve the traffic in both directions.

Figure 2 shows an atypical time-space diagram. It is atypical because, as every traffic engineers knows, life is not so kind as to provide many opportunities of relatively regular arrangements in space and time as those shown in Fig. 2.

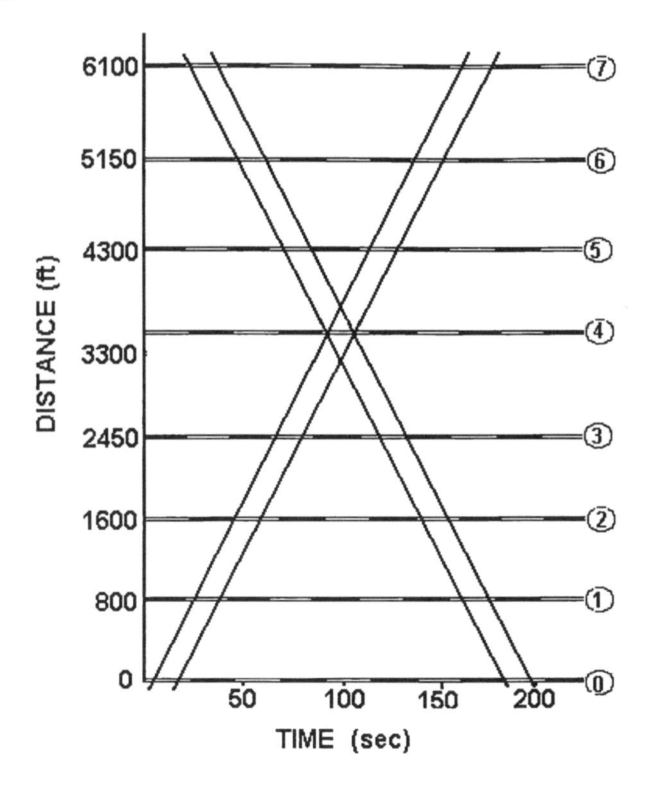

Figure 2. A time-space diagram showing the throughput design of progression.

On the basis of the diagram shown in Fig. 2, we can discuss some of the basic properties of a progression design, which allow variation of a single design in order to accommodate varying demands. Extensive discussions of the principles of progression design can be found in Dunne and Potts (1964), and in Matson et al (1955).

The objective of a progression design is to maximize the green through bands in both directions of an artery. The efficiency of a progression design is defined as the sum of the widths of the two through bands divided by the cycle, indicating the percentage utilization of the cycle. It may be seen that the efficiency of a design does not change if we perform a number of transformations, such as the following:

- We can stretch the time scale of the diagram in Fig. 2 . This amounts to a change from a set of speed and cycle (v, c) to a set $(c/\lambda, \lambda c)$, with λ an arbitrary constant, with the same efficiency of progression.

- We can "shear" the diagram of Fig. 2 from top to bottom, obtaining a progression scheme with different speeds in the two directions and the same efficiency.
- We can trade off some decrease of the bandwidth in one direction for an equal increase in the opposite direction, thus accommodating increasing traffic volumes in the direction getting the increase. Again, the efficiency remains the same.

The stretch of the time scale may be made at some time t_0 corresponding to changing traffic conditions. It is well known in traffic engineering that during light traffic conditions, the shortest possible cycle time is the best. As the traffic volume increases, we want to be able to increase the cycle time, as shown, in Fig. 3. The increase in cycle time corresponds to a decrease in speed, which is anyway a natural consequence of increased traffic volumes.

On the basis of all three of the above observations we can simply seek an optimal design for some speed and cycle, from which we may them obtain a multitude of alternate designs accommodating different traffic demand situations. An optimal design can be obtained from the following observations.

From symmetry, we see that the *offsets* between successive intersections, (i.e. the time differences between the middles of the green phases), must be such that the middle of the green phase of any intersection must coincide with the middle of either the green or the red phase of any other intersection.

If, in addition, all travel times between intersections are integer multiples of the half-cycle, we can achieve a perfect utilization of the green phase at all intersections, and in both directions, resulting in the highest possible efficiency. If the distances between intersections do not allow this perfect

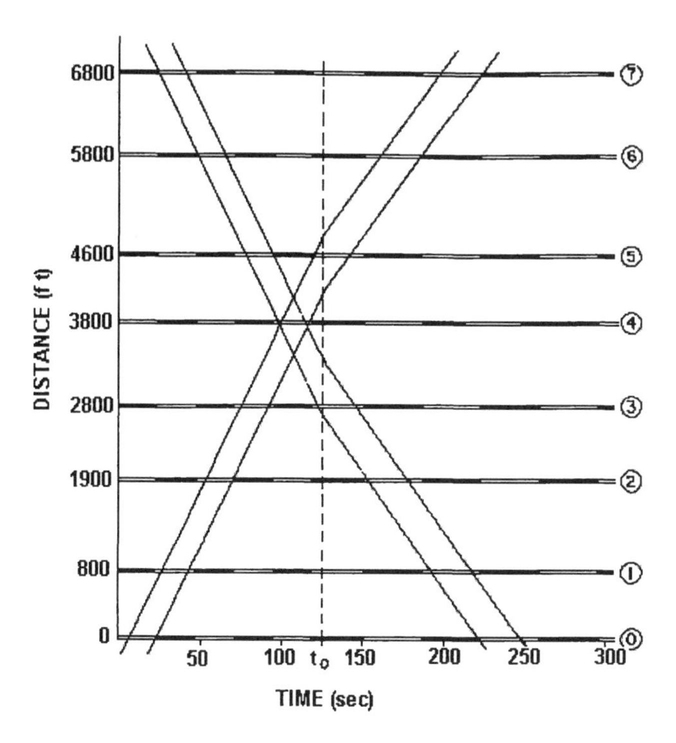

Figure 3. Changing cycle and speed at time t_0 .

solution, we have some *right or left interference* at various intersections, resulting in less than full utilization of the green phases. This is shown in Fig. 4 below, where the case of full utilization of the green phase, diagram (a), is contrasted with the case of partial utilization, diagram (b). Depending on the alignment of mid-phases, some of these intersections produce a right or left interference, or both, in which case they are "critical" intersections. For example in Fig, 2, the critical intersections areNos, 0, 5, 6, and 7. It is at these intersections that adjustments can be made to enlarge the bandwidth of one direction at the expense of the other. In such cases, one can try to improve the system by finding an alignment of midphases such that *the sum of the maximum left interference and the maximum right interference is minimized.* This can be accomplished with the following finite algorithm.

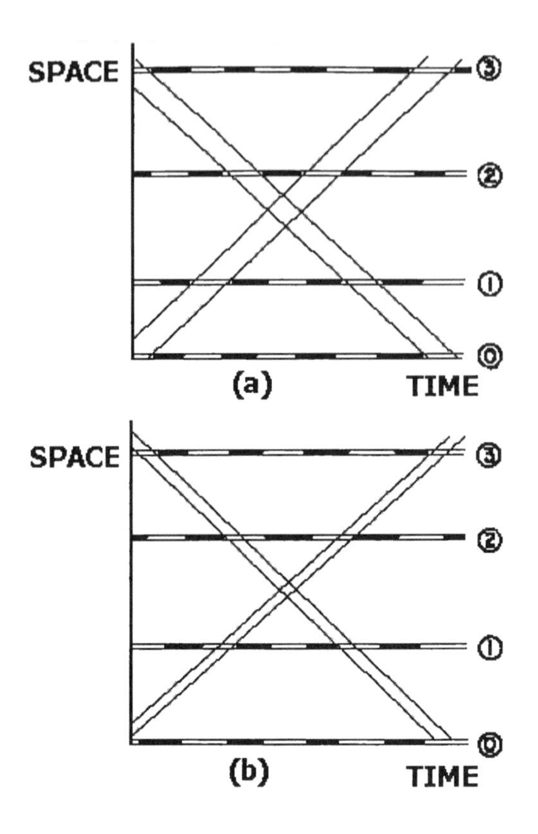

Figure 4. Full or partial utilization of the green phase

1. Let d_i be the distances of the intersections from the one with the minimum green phase, taken as the *reference intersection*, g_i the green phases of these intersections, and g_0 the green phase of the reference intersection. We compute

$$\begin{aligned} t_i &= \frac{d_i}{v}, \\ l_i &= t_i \bmod\left(\frac{c}{2}\right) + \frac{(g_i - g_0)}{2}, \\ r_i &= \frac{c}{2} - l_i + (g_i - g_0), \\ m_i &= \operatorname{int}\frac{t_i}{(c/2)} \qquad i = 1,2,\cdots\cdots,N \quad , \end{aligned} \tag{4}$$

where the symbols "mod" and "int" denote the *modulo* (i.e., the residue after division), and "int" denotes the integer part of the division. The quantities li and r_i are the left and right interferences of the light i on the band emanating from the reference intersection. Aligning the mid-phases results in intersection i being associated with either a right or left interference.

2. We then arrange the left interferences in descending order. Let L_i be this arrangement, and R_i the analogous arrangement for the right interferences. On feasible solution is to select an alignment of mid-phases corresponding to all left interferences, in which the bandwidth is

$$b_0 = g_0 - L_1 \; . \tag{5}$$

3. We can now try to improve the solution by trading off left interferences for right interferences. This is accomplished by shifting the offset of the corresponding intersections by $c/2$. *If* the number of left interferences is k, then the optimum bandwith is obtained by maximizing over k the quantity

$$b_n = g_0 - L_{n+1} - \max_{j=1,n} R_j \; , \qquad n = 1, \cdots, k \; . \tag{6}$$

It should be noted that not all values of k must be considered, but only those corresponding to increasing values of R_j.

4. After we are through with this interim rearrangement of midphases, we align them by computing

$$\begin{aligned} a_i &= (m_i + 1) \bmod 2 \qquad i \le k \; , \\ a_i &= m_i \bmod 2 \qquad k < i \le N \; , \end{aligned} \tag{7}$$

and setting the mid-phases of the intersection i identical to that of the reference intersection when $a_i = 0$, and opposite when $a_i = 1$.

3.3.1 A Mixed Integer Linear Programming Approach

An interesting variation of the progression design was introduced by Little (1964), who allowed small changes of speeds between pairs of intersections, on the theory that drivers could learn to adapt to these changes. Allowing these changes could generally result in wider through-bands.

Little addressed his adjusted progression design problem for an entire network using a mixed-integer linear programming (MILP) approach. The objective function was a weighted sum of the bandwidths of all arteries in the network. Each bandwidth depends linearly on the offsets that are the continuous variables in the MILP formulation. The offsets must satisfy certain constraints involving some integer variables. These constraints express the fact that adding relative offsets between successive pairs of intersections along a closed path in the network, one must obtain an integer multiple of the cycle. In principle, Little's algorithm can handle in a general way any configuration of an arterial network, admittedly by a rather large MILP problem.

If we limit ourselves to maximizing the through-bands along an isolated two-way artery, we have a considerably more tractable problem. We assume that the cycle, splits, and travel time ranges between intersections are given. After we obtain an optimal solution, by shearing the time-space diagram between pairs of intersections, we can obtain one corresponding to equal travel times in both directions. Thus, the solution will be symmetric with respect to the two directions of traffic, although we can subsequently take some bandwidth from one and give it to the other, if the traffic volumes call for it.

Little's formulation starts with the following definitions, where all the variables are expressed as multiples of the cycle c.

$\rho_i =$ red phase of intersection i,

$t_{ij} =$ travel time from i to j $\left(= t_{ji}\right)$,

$\phi_{ij} =$ time from the center of the red phase at i to the center of a red phase at j, where the two red phases are adjacent to a through-band, and on the same side of the band $\left(= \phi_{ij}\right)$,

$w_i\left(\overline{w}_i\right)=$ time from the right (left) side of the red phase of intersection i to the green band, where, from symmetry we need only consider one direction.

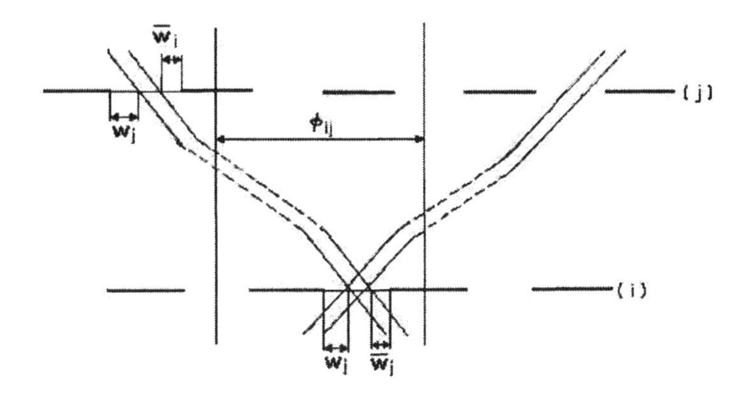

Figure 5. Through-bands and associated parameters in Little's model

From Fig. 5 we obtain the following relationships:

$$\left(\frac{\rho_1}{2}\right)+\overline{w}_i+t_{ij}-\overline{w}_j-\left(\frac{\rho_j}{2}\right)=\frac{m_{ij}}{2} \quad ,$$
$$\left(\frac{\rho_j}{2}\right)+w_j+t_{ij}-w_i-\left(\frac{\rho_i}{2}\right)=\frac{m_{ij}}{2} \quad , \qquad (8)$$

where

$$m_{ij}=2\phi_{ij}=\text{integer} . \qquad (9)$$

In addition, looking at Fig. 4 , we obtain the constraints

$$w_i + b \le 1 - \rho_i \ ,$$
$$\overline{w}_i + b \le 1 - \rho_i \ . \tag{10}$$

The optimization objective can now be defined as follows: Find b, $w_i, \overline{w}_i$, and m_{ij} which maximize b subject to the constraints expressed by Eqs. 8 to 10, plus the positivity constraint

$$b, w_i, \overline{w}_i \ge 0 \ . \tag{11}$$

The optimization problem is now a MILP problem, with the constraint that m_{ij} be integer p-providing the "integer" part of the MILP problem.

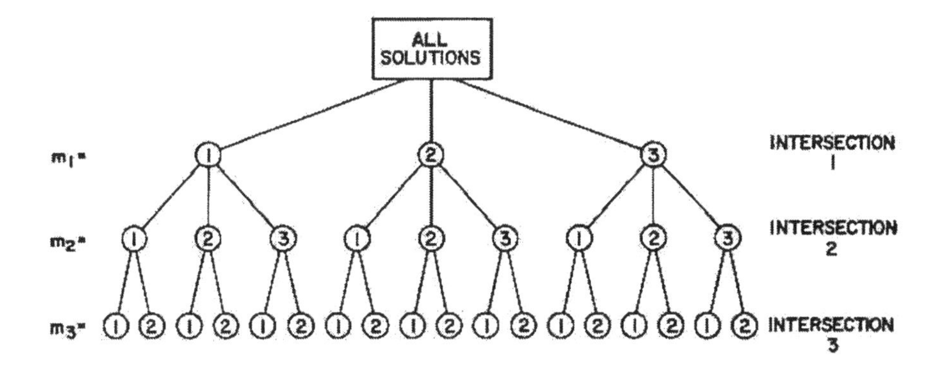

Figure 6. The synchronization "tree" used by Little in his branch and bound algorithm

As in all MILP problems, for any set of the integer values, the problem is a linear Programming (LP) one. For the MILP problem, Little suggested a *branch and bound* approach, which provides a "tree" one searches on the way toward the solution, such as that shown in Fig. 5. The tree is formed by considering intersections one at a time, and all the values of the integer variable m associated with the pair of the intersection at hand and a reference intersection. For each node of the tree, we solve a LP problem to obtain a value of the through-band, assuming fixed values of m_{ij}, for a subset of the intersections. The branch and bound technique starts with the node with the highest through-band and searches along the branches of the tree, until all the values of m_{ij} have been selected.

The advantage of Little's method is that it can be extended to consider various other possibilities regarding light synchronization, such as the following:

- The change of speeds between successive pairs of intersections may be allowed to vary between two bounds. This is accomplished by introducing additional variables t_{ij} which must satisfy appropriate constraints.
- The through-bands in the two directions may be made to have any desired ratio.
- The cycle may vary between two bounds.
- The method may be adapted for networks of arteries, by introducing "loop constraints", which must be satisfied by variables such as the w_{ii} along closed loops of the network. Little's suggestion for an objective function of this problem was a weighted sum of the through-bands along all the arteries of the network.

Little's method, which also came to be known as the *MAXBAND* method, has been extended by various investigators in order to consider a variety of new aspect of the problem, such as: time of clearance of existing queue, left-turn movements, and different bandwidths for each link (*MULTIBAND*) for each link of the arterial, as discussed by Gartner (1991), and Gartner et al (1991).

3.3.2 The effect of queues on progression

All the synchronization schemes discussed above have one basic deficiency, namely, they are valid when the streets have rather light traffic. As everyone knows, when the traffic density increases, queues of cars bock the path of following cars through successive green lights, and the progression scheme becomes meaningless. This situation has been investigated by Gazis (1965) and is shown in Fig. 7. If there are no queues, the ideal offset between intersections N and M would be such that the green phase of intersection M starts d/v seconds after that of intersection N, where d is the distance of the two intersections, and v the speed of a moving platoon of cars.

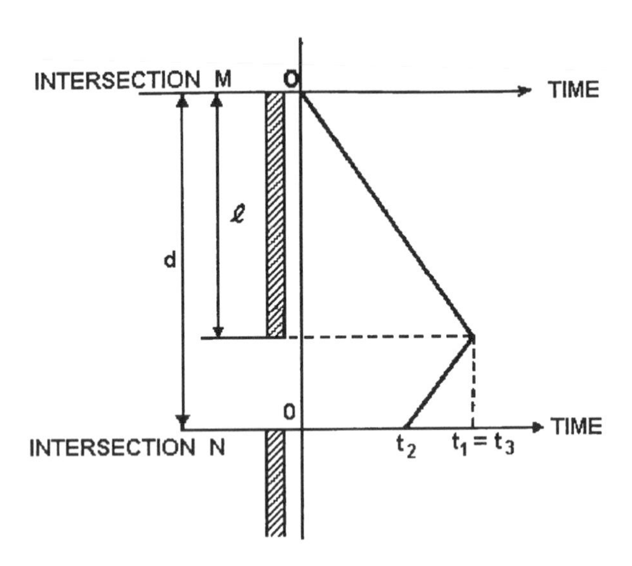

Figure 7. The effect of a queue on progression design

If, however, there is queue of length l stopped at intersection M, then the onset of the green phase at M induces a "starting wave" that moves toward the end of the queue with speed v_1, reaching the end of the queue after time l/v_1. If the leader of the platoon starting from intersection N at the beginning of its green phase is to reach the end of the queue in front just as its last cars starts moving, then the car must start at time t_2, measured from the start of the green phase at M, where t_2 is given by

$$t_2 = \frac{l}{v_1} - \frac{d-l}{v} \quad . \tag{12}$$

If the queue stopped at N is released by turning the light green at time t_2, the front of the platoon, traveling at speed v, will reach the end of the queue at time

$$t_3 = t_2 + \frac{d-l}{v} \tag{13}$$

Setting $t_1 = t_3$, we obtain a situation where the interference between platoons is more or less minimized.

It is seen from Eq. 12 that t_1 is positive when l is sufficiently large, leading to a situation which has been referred to by some as "reverse progression". Whatever the appellation, the point is that adjustment must be made for the existence of a queue, if the upstream platoon is to go through intersection M without stopping. It may very well be that in extreme conditions, not all of the platoon cars will clear intersection, but some will start a new queue, but this is the best that can be done under the circumstances.

Now, one may suggest a method for finding the relative offset of the two lights for all traffic situations, from light traffic to heavy traffic, (but not extreme congestion). When the traffic is light, l is generally zero, and the relative offset is

$$t_2 = -\frac{d}{v}, \tag{14}$$

the standard progression offset. As l increases, t_2 also increases, up to the point of becoming equal to

$$t_2 = t_1 = t_3 = \frac{d}{v_1} \tag{15}$$

In this case $l = d$, and we have reached the level of full saturation. If, along an artery, the density builds up gradually from A to B (Fig. 8), and then decreases toward zero from B to C, then the offset with respect to A may be given by a curve such as $\phi(x)$, which can be obtained following the method given above.

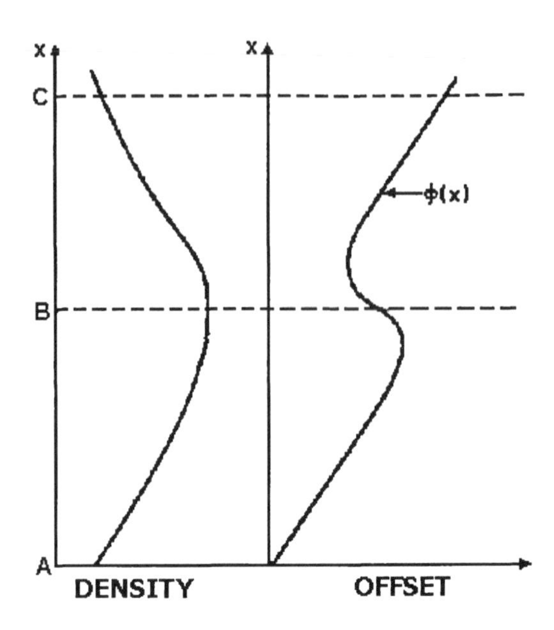

Figure 8. Offset with respect to the origin A when queue lengths

It should be pointed out that such a design may only indirectly reduce delays, by preventing or reducing the blocking of intersections, and the infamous "gridlock". Its main effect would be to minimize the number of starts and stops of vehicles by starting each vehicle only when there is room enough for it to proceed.

3.3.3 The TRANSYT method

Adjustments for the existence of queues has been done, directly of indirectly, by many investigators. One of the most notable examples is the TRANSYT method of Robertson (1969). TRANSYT simulates the motion of cars through a network, evaluating an objective function which measures the "inconvenience to drivers", and then finds a good set of offsets by searching over the domain of allowable offsets. TRANSYT is probably the best known, and most frequently used road control strategy. Field implementation of TRANSYT designs have indicated a possible reduction of up to 16% in travel times through a network.

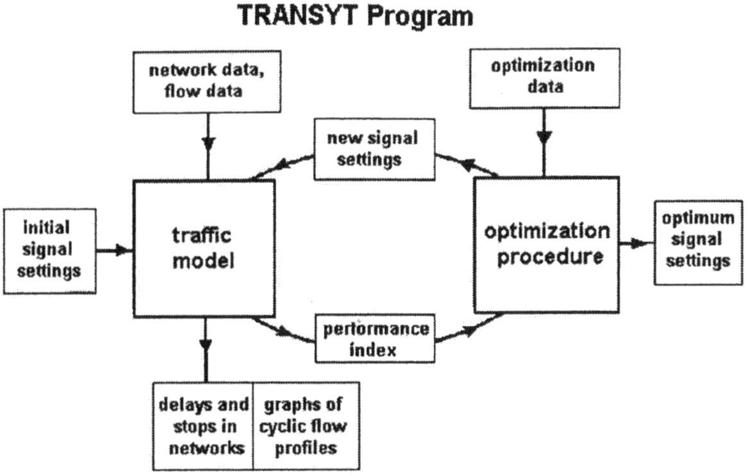

Figure 9. The structure of the TRANSYT program

Figure 9 shows the basic structure of TRANSYT, as described by Robertson. The "initial signal settings" include some pre-specified coordination of lights, the minimum green durations for each intersection, and the initial choice of splits, offsets, and cycle time, the latter assumed to be the same for all network intersections. The "network data, flow data" comprise the network geometry, the saturation flows, the link travel times, the constant and known rates of turning movements in each intersections, and some known, constant demands. The "traffic model" consists of nodes (intersections) and links (connecting streets). Cars move along the links of the system in platoons, with some platoon dispersion taking place as the platoon moves along a link. This platoon dispersion is the change in the spatial distribution of a group of vehicles due to the difference in their speeds. TRANSYT also allows some deviation from fixed splits. Thus, in addition to optimizing the system over the allowable ranges of offsets, it tries to improve the traffic further by varying the allocation of green in each intersection.

The TRANSYT procedure comprises a model used to calculate delays and stops, and a "hill-climbing" technique for minimizing an objective function, D, given by

$$D = \sum_{i=1}^{N} \left(d_i + K h_i \right) \quad , \tag{16}$$

where d_i is the average delay per hour on the ith link of a network of N links, h_i is the average number of stops per second on this link, and K is a weighting factor.

The model of traffic movement consists of generating a pattern of flow on network links by manipulating three types of flow patterns for each direction of flow, namely,

1. The IN pattern, which is the flow of traffic past the downstream end of a link, if traffic is not impeded by a traffic light.
2. The OUT pattern, which is the actual traffic flow past the intersection into the next link, and
3. The GO pattern, which is the flow obtained if there is a residual queue upstream of the intersection during the entire green phase.

The OUT pattern of one intersection generates the IN pattern of the next downstream intersection, with some allowance of "dispersion", the spreading out of platoons due to varying speed between platoon cars. A typical IN pattern is shown in Fig. 10, in the form of a histogram of flow during each of 50 equal fractions of the cycle.

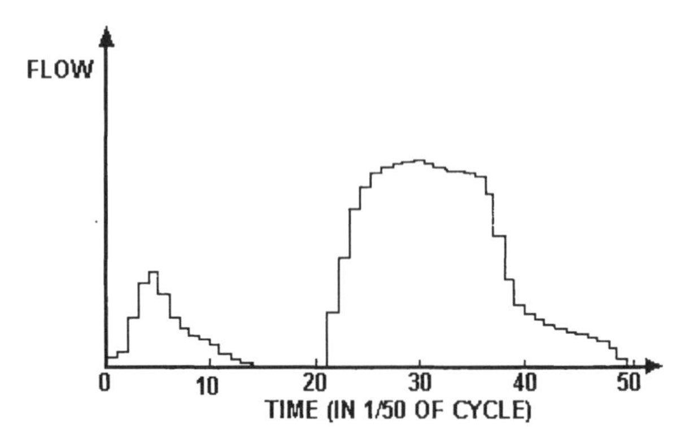

Figure 10. A typical IN flow pattern of the TRANSYT program

It may be seen that the position of the green phase along the time axis of the IN pattern allows one to compute both the delay and the number of stops at an intersection, which contribute to the objective function D. A minimum for D is then obtained by the following "hill-climbing" procedure. The offset and/or split of a single intersection is changed by a small amount, and a new value of D is computed. The change of offset and split is continued until a (local) minimum of D is obtained. The process is repeated by varying the offset and split of all other intersections in pre-selected sequence, completing one iteration. The procedure is stopped after several iterations, when the marginal improvement of the objective function begins to get very small.

3.3.4 The Combination Method

Another method of signal synchronization is the *Combination Method*, proposed by Hillier and Lott (1966). The Combination Method assumes a common cycle and given split at each intersection. In addition, it is assumed that

1. The amount of traffic at each intersection does not depend on the traffic light settings, and
2. The delay along a given link of a network depends only on the settings of the lights at the two ends of the link.

The method is based on the observation that, if the above assumptions are true, links can be combined in series or in parallel, in computing and minimizing delay. Consider a pair of intersections A and B (Fig. 11), connected by two traffic links. Assumption 2 above states that the delay of traffic along link 1 depends only on the offset between A and B. Let this delay be given by the function

$$D_1^{AB}(z_{AB}) \qquad 0 \le z_{AB} \le c \ , \tag{17}$$

where z_{AB} is the offset of B with respect to A. The quantity D_1^{AB} is defined within the interval $(0,c)$, where c is the cycle, since it is a periodic function of z with period c.

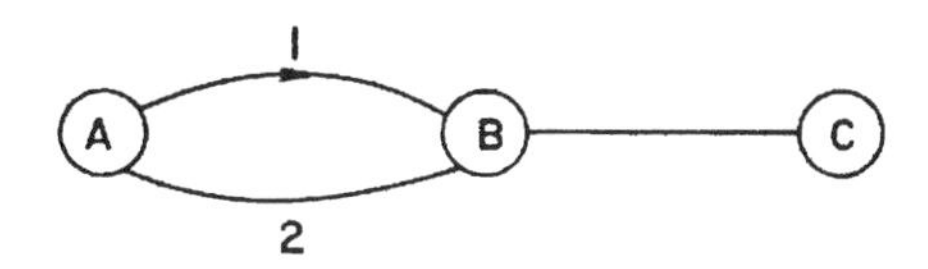

Figure 11. Link combination in parallel and in series

We also define a similar function $D_2^{AB}(z_{AB})$ for the delay along link 2. Since it is assumed that the two delays are independent of each other, they can be added in computing the total delay for the network which includes A and B. We can then replace the two links by a single one associated with a delay function

$$D^{AB}(z_{AB}) = D_1^{AB}(z_{AB}) + D_2^{AB}(z_{AB}) \tag{18}$$

We next consider two links in series connecting three consecutive intersections A, B, and C. We define the delay functions

$$\begin{aligned} &D^{AB}(z_{AB}) \qquad 0 \le z_{AB} \le c \ , \\ &D^{BC}(z_{BC}) \qquad 0 \le z_{BC} \le c \ . \end{aligned} \tag{19}$$

We then compute the delay of traffic moving from A to C, D^{AC}, as a function of the offset of the light at C with respect to the light at A, z_{AC}. This delay depends, of course on the offset of the light B as well. For any value of z_{AC}, we pick the offset at B so as to obtain the minimum delay from A to C. We do this by defining

$$D^{AC}(z_{AC}) = \min_{0 \le z_{AB} \le c} \left[D^{AB}(z_{AB}) + D^{BC}(\tilde{z}_{BC}) \right] \tag{20}$$

where

$$\begin{aligned} \tilde{z}_{BC} &= z_{AC} - z_{AB} \qquad && if \quad z_{AC} \ge z_{AB} \\ &= c + z_{AC} - z_{AB} \qquad && if \quad z_{AC} < z_{AB} \ . \end{aligned} \tag{21}$$

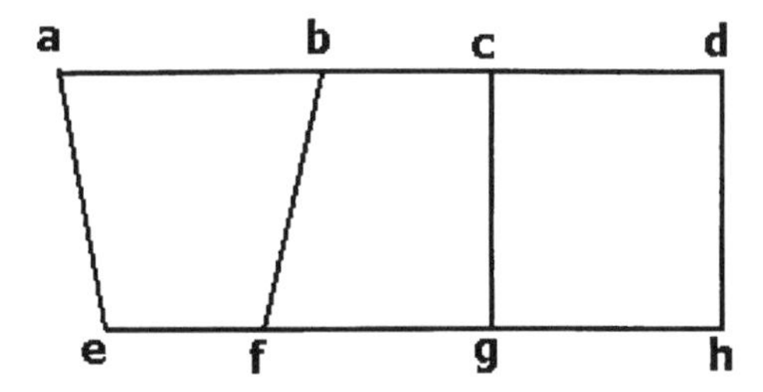

Figure 12. A "ladder" network which can be reduced to a single link

This allows us to replace the links AB and BC by a single link AC in computing the total delay for the network.

Combining links in series and in parallel yields a reduction of the network to one with fewer links. Hillier observed that a network in the form of a ladder, such as that shown in Fig. 12, could be reduced to a single link, simplifying the optimization process. Allsop (1968a, 1968b) has shown that any network cab be "condensed" by the above process into an irreducible form, in which at least three links meet at each node. For example, the network shown in Fig. 13 (a) can be condensed into that shown in Fig. 13 (b) by combining the links as follows:

AB and BC in series into AC ,
AC and CD in series into $(AD)_1$,
$(AD)_1$ and AD in parallel into $(AD)_2$,
$(AD)_2$ and DE in series into AE ,
EF and FG in series into EG ,
GH and HI in series into GI ,
IJ and JA in series into IA .

In combining links in series, Hillier suggested discretizing the interval $0 \leq z \leq c$, and then searching for the minimum of Eq. 20 at discrete values of z .

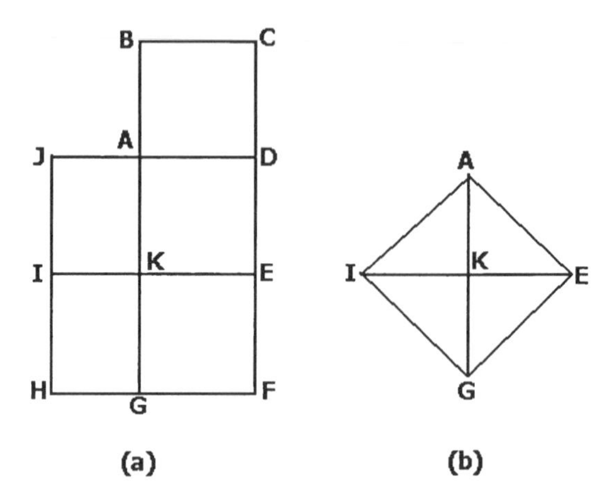

Figure 13. Reduction of network (a) to its irreducible form (b)

The synchronization of lights for an irreducible network such as that of Fig. 13 (b) was also considered by Allsop (1968a, 1968b). He proposed an iterative scheme for minimizing the delay in such a network by an appropriate selection of offsets. The scheme starts with a single link, and proceeds to form larger and larger sub-networks in successive stages. At each stage, one estimates, for any combination of offsets at intersections on the boundary of the sub-network, a set of offsets for the internal intersections, which minimizes delay. The process proceeds until all links are included, resulting in the solution to the synchronization problem.

3.3.5 The SIGOP Program

The SIGOP (Signal Optimization) program (SIGOP 1966) is one developed under the auspices of the U. S. Bureau of Public Roads by the Traffic Research Corporation. The program tries to minimize, in a least-square sense, the deviation of the relative offsets of neighboring intersections from a set of "ideal" offsets, which are stipulated, computed or adjusted on the basis of real life measurements. It is implicitly assumed that the delay is a periodic function of the relative offsets, as shown in Fig. 14 (a), with a minimum at some value between $-c/2$ and $c/2$, where c is the cycle, or this value incremented by a multiple of c. It is further assumed that this ideal relative offset is independent of the offsets of the other intersections. This delay function may be approximated by a set of parabolas with the same minimum, as shown in Fig. 14 (b).

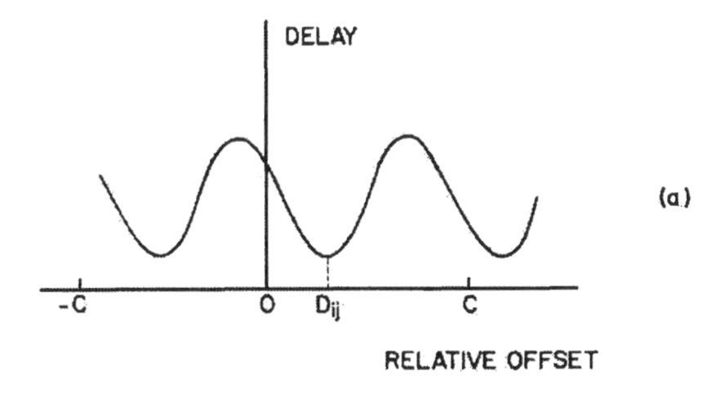

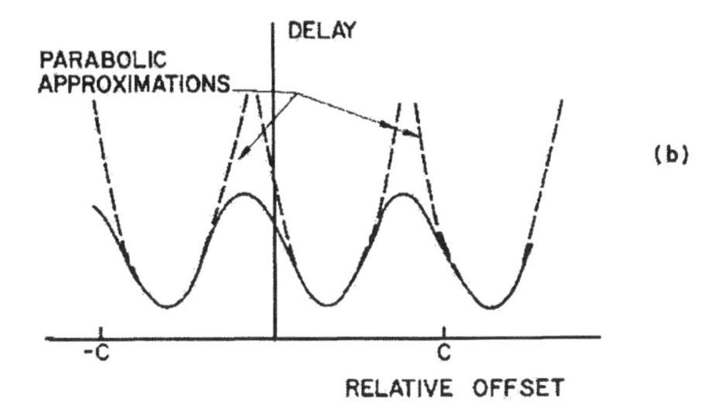

Figure 14. Delay functions used in the SIGOP program

This leads to an objective function of the type

$$D = \sum_{ij} \alpha_{ij} \left(D_{ij} + \theta_i - \theta_j + M_{ij} c \right)^2 , \tag{22}$$

where D_{ij} is the ideal relative offset between intersections i and j, θ_j is the design offset of intersection j, α_{ij} is a "link importance factor", weighing the traffic from i to j, and M_{ij} is an integer value, usually 0 or ± 1, which is selected so that the relationship

$$\frac{-c}{2} \leq D_{ij} + \theta_i - \theta_j + M_{ij} c \leq \frac{c}{2} \tag{23}$$

is satisfied. Now, the procedure for minimizing D is a combination of a Monte Carlo random starting point routine, and a direct search procedure. In effect, we divide the multi-dimensional domain into different regions

$$0 \le \theta_j \le c \tag{24}$$

corresponding to various choices of the values of M_{ij} which satisfy Eq. 23. The SIGOP program searches for the global minimum of D within the domain defined by Eq. 24, by obtaining systematically local minima through the following sequence of steps:

1. A random choice is made of a θ_j which satisfies Eq. 24, and the corresponding region is determined by finding all the M_{ij} which satisfy Eq. 23.
2. The minimum of D corresponding to the region determined in step 1 is found by diferentiating Eq. 22 and solving the resulting system of linear equations in θ_j.
3. Step 2 is repeated, changing each M_{ij} by ± 1, until a local minimum is obtained. This is so if a further change of any M_{ij} by ± 1 would yield a higher value of the minimum obtained through step 2.

Steps 1 to 3 constitute one "Monte Carlo game". We then start mew games with step 1 and obtain new values for local minima, until we are reasonably certain that the global minimum is among all the local minima obtained through the set of Monte Carlo games. Such a certainty that one has not missed the global minimum is a mater of experience and judgement, which is a requirement for proper use of the SIGOP program.

It may be pointed out that an exhaustive search of **all** the regions of the domain of Eq. 24, for all possible values of M_{ij}, runs into an extraordinary amount of computations. For example, it has been estimated by Ross (1972) that a square grid with 25 signalized intersections, all serving two-way streets, would correspond to about 10^{28} regions of the domain defined by Eq. 24.

3.3.6 Additional contributions to the synchronization problem

There have been numerous additional contributions to the study of the signal synchronization problem. In fact, this subject has been one of the most investigated topics in traffic theory.

Buckley et al (1966) have contributed methods for maximizing through-bands for regular rectangular grids of one-way arteries. Bavarez and Newell (1967) contributed an optimization scheme for through-bands for one-way arteries, for a variety of objective functions involving both delay to drivers and number of stops. In effect, Bavarez and Newell start a platoon at the beginning of an artery, and look for a synchronization scheme that does the best possible job of serving this platoon. According to their analysis, for any time dependence of the traffic input at the entrance of the artery, there exists in general a synchronization scheme better than the one corresponding to a continuous through-band. The optimum scheme is obtained by "juggling" the offsets of the intersections near the entrance of the artery, deviating from the continuous through-band design. Of course, the method assumes exact knowledge of the time dependence of the traffic input into the artery, which is not always available. Gartner (1972a) used arguments of graph theory to prove theorems concerning the optimal through-band design for a network. The theorems involve relationships between design parameters such as offsets, and are likened by Gartner to Kirchhoff's laws for electrical networks.

Two other contributions, by Okutani (1972) and by Gartner (1972b) use dynamic programming for optimizing the synchronization of traffic signals, by minimizing delay an/or number of stops.

Okutani assumes that the delay between pairs of intersections depends only on the relative offsets of these intersections . He then uses dynamic programming to obtain the optimum synchronization, by dividing the network into sections, which are treated as stages in a multi-stage decision process. He applied Bellman's principle of optimality to obtain a recursion relationship between the optimum design for n sections, and for $n + 1$ sections. Okutani also suggested an alternative approach to the problem based on Pontryagin's maximum principle (see Katz 1962).

Gartner considers portions of a network as stages in a multi-stage decision process. Each stage is small enough, so that an exhaustive search on a small number of independent variables is possible in optimizing the stage. The global optimization consists of minimizing a suitable objective function associated with the delay and/or the number of stops of the traffic. The multi-stage process is optimized using dynamic programming. Gartner also suggests a procedure, based on graph-theoretical concepts, which tends to minimize the required computational effort.

The preceding methods for optimizing the synchronization of lights in a network have been tested, to some extent, in real-life situations. These tests have shown that schemes, which take into account the formation of queues at intersections are distinctly better than those that simply maximize a through-band, insofar as minimizing delay is concerned. The margin of improvement over simple through-band maximization was found to be as high as 15%, for moderately heavy traffic, but **not** oversaturated conditions. In addition, taking into account the existence of a queue at the beginning of the green phase tends to mitigate the danger of the familiar "gridlock" produced by cars entering into an intersection, without being able to clear the intersection before the cross-traffic green starts. In general, designs that take into account of the existence of queues tend to improve performance during periods of moderate to heavy traffic, but **not** congestion.

3.4 TRAFFIC RESPONSIVE OPERATION OF TRAFFIC LIGHTS

The operation of traffic lights evolved from fixed-time operation to "traffic actuated" operation, based on actual real-time measurements of traffic along competing directions. Typically, such actuation involved extending the green phase, up to a maximum, to accommodate an increasing demand of a given direction. When two competing directions showed signs of traffic demand increasing toward congestion, both directions would get the maximum green. This is desirable in conditions of heavy traffic, because it minimized the effect of the lost time for clearance of the intersection before switching. When computers were introduced for traffic control, in the 1960s, many of these heuristic schemes found heir way into the code which generated the computer decisions regarding the settings of the traffic lights. However, the schemes did not necessarily involve a thorough analysis of the performance of systems guided by them, because the state of the art of such an analysis was not up to the task. Several contributions to the state of the art have been made since that time, and will be discussed in the sequel.

3.4.1 Single Intersection

One of the first contributions in the area of traffic responsive signals for a single intersection was made by Dunne and Potts (1964), who proposed their own heuristic algorithm for traffic actuation, and also gave an interesting method for representing and analyzing the operation of traffic lights. The representation consists of describing the condition of the intersection in the "state space" (x_1, x_2), where x_1 and x_2 are the sizes of queues waiting to be served, along the two competing directions of the intersection, as shown in Fig. 15. Assuming that the variables x_i are continuous, the condition of the intersection will be represented by a continuous line. If, in addition, the traffic inputs are assumed to be constant and continuous, and the flow rates during the green phase equal to the saturation flows, then the condition of the intersection is represented by a point moving along one of three possible straight lines:

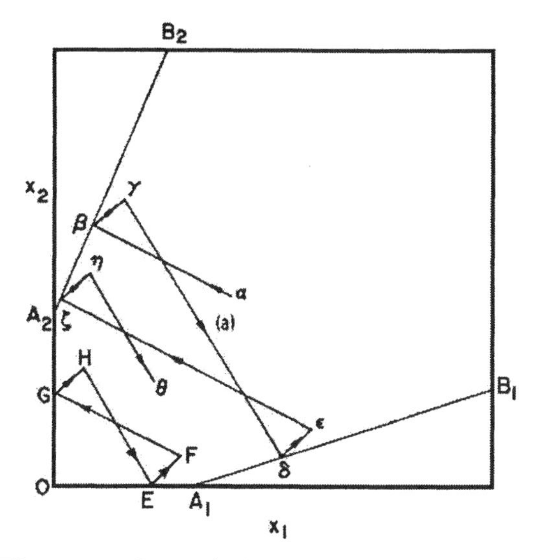

Figure 15. Queue behavior at an intersection

1. When the light favors direction 1, the state of the intersection, (x_1, x_2), moves along the direction of the vector $(q_1 - s_1, q_2,)$, where q_i are the input rates and s_i the saturation flows along the two competing directions.
2. When the light favors direction 2, the state of the intersection moves along the vector $(q_1, q_2 - s_2)$.
3. During the period of clearance which includes the yellow phase of the light, the state of the intersection moves along the direction of thevector (q_1, q_2), as both queues are increasing.

In cases 1 and 2, it is assumed that there is a queue of cars waiting to be served at the start of the green phase, for both directions. When the queue in the favored direction is fully served, the corresponding axis $x_i = 0$ acts as a *barrier* for the state of the intersection, and the state of the intersection moves along this barrier until the light is changed, with only the not favored queue increasing.

Dunne and Potts went on to propose and test an algorithm which treats the lines OA_1B_1 and OA_2B_2 in Fig. 15 as "reflecting barriers", meaning that the light changes as soon as the state of the intersection reaches either one of these lines attempting to cross. Incidentally when the reflecting barriers are just the axes $x_i = 0$, the Dunne-Potts algorithm treats the case known as the *Saturation Flow Algorithm*, in which the light changes only when the favored queue is exhausted, which may be impractical in many real-life situations.

The Dunne-Potts algorithm yields a trajectory in state space such as the line $\alpha\beta\gamma\delta\varepsilon\zeta\eta\theta$ in Fig. 15. Under appropriate conditions, it may yield a "limit cycle", such as that represented by the line $EFGH$ in Fig. 15. The algorithm proceeds as follows:

The algorithm is *stable*, that is, it yields bounded values of $x_i(\infty)$, when the intersection is undersaturated, i.e. when

$$\frac{q_1}{s_1} + \frac{q_2}{s_2} < 1 \ . \tag{25}$$

In that case, a limit cycle exists, with phases $g_{1\infty}$ and $g_{2\infty}$ given by

$$g_{i\infty} = \frac{2ly_i}{1-Y} \qquad i = 1,2, \tag{26}$$

where

$$y_i = \frac{q_i}{s_i} \quad i = 1,2, \qquad Y = y_1 + y_2 \ , \tag{27}$$

and l is the lost time for clearance at phase change-over. The duration of the limit cycle is

$$c = g_{1\infty} + g_{2\infty} + 2l = \frac{2l}{1-Y} \quad . \tag{28}$$

The above results concerning the limit cycle are applicable, whether or not the portions $A_i B_i$ of the reflecting barriers coincide with the axes x_i . However, Dunne and Potts point out that the delay to users is minimal when the barriers $A_i B_{ii}$ coincide with axes x_i , which is the case of the saturation flow algorithm.

Another contribution to the optimization of the operation of a traffic-actuated traffic light was that of Grafton and Newell (1967). They defined the saturation flow operation of the traffic light as the *basis policy*, and they compared it with two other *comparison* policies. One of them allowed switching of a traffic light before the favored queue was empty, and the other allowed an extension of the green phase past the time when the favored queue became empty. Comparing the basis policy with the comparison policies led to the conclusion that there were domains in the x_i space where the basis policy was optimal , and other domains in which the comparison policies were optimal. Grafton and Newell then applied Bellman's *Principle of Optimality* (Bellman and Dreyfus, 1962) to complete the solution. In this case Bellman's principle expressed the "recursive property" of the optimal solution, namely, that whatever the initial state of the system and the initial decisions up to an intermediate time τ , the remaining decisions after time τ should lead to an optimal policy with respect to the state resulting from the first decisions.

Another interesting contribution in this area is that of van Zijverden and Kwakernaak (1969), who gave a general approach for a good sub-optimal control of a complex intersection serving time-varying demands. They applied principles of optimization of stochastic control systems given by Kwakernaak (1966), and obtained an optimal solution through an exhaustive search from among all possible sequence of light settings over a short time horizon, and repetition of this process thereafter.

3.4.2 Systems of Intersections

A number of coordinated Traffic-responsive strategies for networks of arteries have been developed, and frequently applied to various systems around the world. One of the first schemes for a fully adaptive control of a large number of intersections was given by Miller (1965). His algorithm involves making a binary decision at time intervals h, for every traffic light, whether to leave the light in its present state or change it. The decision is based on the computation of the delay for the two options.

As an example, let us assume that at a N-S, E-W intersection the traffic light is green in the N-S direction. If we keep the light in this state for an extra h seconds, the reduction of delay for the N-S traffic can be found to be

$$\Delta D_{NS} = (a + r_{NS} + l_{NS})\left\{\delta_N + \delta_S - q_N \frac{1-(\delta_N / s_N)}{1-(q_N / s_N)} - q_S \frac{1-(\delta_S / s_S)}{1-(q_S / s_S)}\right\} \tag{29}$$

where

a is the duration of the amber (yellow) phase,
l_{NS} is the lost time for clearance between favoring the N-S or E-W direction,
δ_N , δ_S are the number of vehicles expected to move in the N and S directions during h seconds,
q_N , q_S are the arrival rates of vehicles per h seconds in the N and S directions, and
s_N , s_S are the corresponding saturation flows in the two directions.

If we defer the start of the E-W green phase by h seconds, we impose some additional delay to the vehicles already there, and those that may arrive before the queues in the E-W directions are exhausted, when we do give them the green light. This additional delay is estimated to be

$$\Delta D_{EW} = h\left(n_W + n_s + \sum_{i=1}^{k_w} q_W + \sum_{i=1}^{k_E} q_E \right) \tag{30}$$

where n_W, n_E are the vehicles already queued in the W and E directions, q_W, q_E are the corresponding arrival rates, and k_W, k_E are the smaller integers such that

$$n_W + \sum_{i=1}^{k_W} q_W - \sum_{i=2+l_W/h}^{k_W} s_W \leq 0 \, ,$$
$$n_E + \sum_{i=1}^{k_E} q_E - \sum_{i=2+l_E/h}^{k_E} s_E \leq 0 \, . \tag{31}$$

where s_W, s_E are the saturation flows in the W, E directions per h seconds. and l_W, l_E are the lost times in these directions. Equations 31 determine the number of intervals of h seconds after the beginning of the green phase, during which there is still a residual queue. Combining Eqs. 30 and 31, we find that, if we extend the N-S green phase by h seconds, the net reduction of delay to all vehicles around the intersection is

$$\Delta D = \Delta D_{NS} - \Delta D_{EW} \, . \tag{32}$$

If $\Delta D > 0$, it pays to leave the signal favoring the N-S traffic for h , repeating the process thereafter, as long as the maximum green phase allowable is not exceeded.

3.4.3 Additional contributions to Synchronization of Intersection Networks

A somewhat similar approach to Miller's was used by Ross et al (1970), for a small group of intersection around a "critical intersection", which is most likely to cause the highest delays. Both Miller and Ross et al tested their strategies by simulation, and found them to be beneficial, particularly in accommodating large fluctuations of traffic such as those caused by temporary interruption of flow. Improvements to both approaches could be made by using the familiar adaptive control technique of a *Rolling Horizon*, namely, predicting for the next, say, five periods and applying the solution for one period.

A control strategy involving small adjustments about a progression design, which do not extend from one cycle into the next, was tested in Glasgow, as discussed by Holroyd and Hillier (1969). The strategy, known as EQUISAT, consisted of equalizing saturation flows along the two competing directions of an intersection, leading to the relationship

$$\frac{q_1}{s_1 g_2} = \frac{q_2}{s_2 g_1} , \tag{33}$$

where $q_i, s_i,$ and g_i, $(i = 1,2)$, are the traffic input rates, saturation flows and green phases, respectively.

Another strategy was proposed by Koshi (1972) for the city of Tokyo, and it is one of the most adaptive ones proposed for a relatively large traffic system. In general, for large systems, the general approach has been to separate control of intersections into "Macrocontrol" and "Microcontrol". Macrocontrol comprises an overall strategy for the entire system, and microcontrol involves small changes in this strategy to accommodate detectable departures from expected demand at some "critical intersections". In effect, Koshi introduces microcontrol over the entire system, but he does it in a way that generates a continuously changing grand strategy for the entire system. The optimization is performed over the time period of one cycle. Koshi looks at every intersection and tests to see whether or not a small advance or delay of the next traffic light phase may decrease the aggregate delay to the users. If it does, the change in the offset is implemented, resulting in a continuous application of a "policy improvement" scheme, which is responsive to changing traffic conditions. It should be mentioned here that a continuous change from one grand strategy to another has distinct advantages over an abrupt one, which may cause problems in transition from one to another.

A number of other programs have been contributed over the years for improving the control of traffic lights in a network. The onset of the use of computers for real-time traffic control, starting in the 1960s, has made the application of such contributions to real systems possible. Among these contributions are the following:

a) The SCOOT Program

The SCOOT program was first developed by Hunt et al (1982), and has been extended in several respects by others. It may be viewed as the traffic-responsive version of TRANSYT, which was discussed earlier. It has been used in many cities around the world, particularly in the United Kingdom.

SCOOT, like other systems, utilizes measurements of traffic volume and "occupancy", which is the percentage of the time a sensor detects a passing vehicle, and hence is related to the density. The measurements are taken at the entrance of a link. SCOOT runs on a central computer, and uses an approacch similar to that of TRANSYT for minimizing delays. The main difference is that, instead of assuming historically available fixed traffic inputs like TRANSYT, SCOOT uses real-time measurements of traffic, and is run repeatedly in real time to investigate the effects of incremental changes of splits, offsets, and cycle time. After defining a "performance index" related to delays, SCOOT selects the changes which would result in an improvement of the performance index, and runs the traffic lights accordingly. Results of field evaluations have been discussed by Luk (1984).

b) The OPAC, PRODYN, CRONOS, and COP Programs

Advanced traffic-responsive programs for networks include OPAC (Gartner, 1983), PRODYN (Farges et al, (1983), CRONOS (Boillot et al, 1992) and COP (Sen and Head, 1997). These strategies do not consider explicitly splits, offsets, or cycles of traffic lights. On the basis of some pre-specified staging, they try to find, in real-time, the optimal specification of the next few switching times, $\tau_i, i = 1,2,...,$ over a future time horizon H, starting from the current time t and the currently applied stage. In order to obtain the optimal switching times, these methods solve dynamic optimization problems using a variety of realistic, dynamic traffic models, which include some binary variables reflecting the impact of red/green phases on traffic flow and the associated performance index. Some constraints, such as constraints on maximum and minimum splits, are also included. The performance index which is to be minimized is the total time spent by all vehicles during the time horizon H.

Recognizing that it is difficult to predict traffic accurately for long periods, these programs use the concept of a *Rolling Horizon*, H, but they apply the results of optimization to a much shorter period, after which new data are collected and the optimization program is solved again for a time horizon H.

One basic problem with these methods is the introduction of binary variables, which require exponential-complexity algorithms for global optimization. Specifically, OPAC employs complete enumeration of choices corresponding to integer switching times, while PRODYN and COP employ dynamic programming. Because of the exponential complexity of these methods, the resulting solutions, although conceptually applicable to an entire network, cannot be easily applied to more than one intersection. Therefore, the optimization is generally limited to optimizing single intersections, whose settings may be coordinated by a superimposed heuristic control strategy, as discussed by Kessaki et al (1990). CRONOS differs from the other methods, in that it employs a heuristic global optimization method with polynomial complexity, which permits consideration of several intersections, although again intersections are optimized individually.

3.4.4 The Onset of Oversaturation

Oversaturated systems are discussed in detail in the following section. Here, we shall make some comments regarding the behavior, and possible improvements, of a single artery whose traffic movement may be controlled by a series of traffic lights, as discussed by Gazis (1965). It is well known that such arteries, with traffic light operation designed for progression, fail miserably during periods of congestion. The reason is shown in Fig. 16, where, for convenience, traffic characteristics are drawn as continuous lines

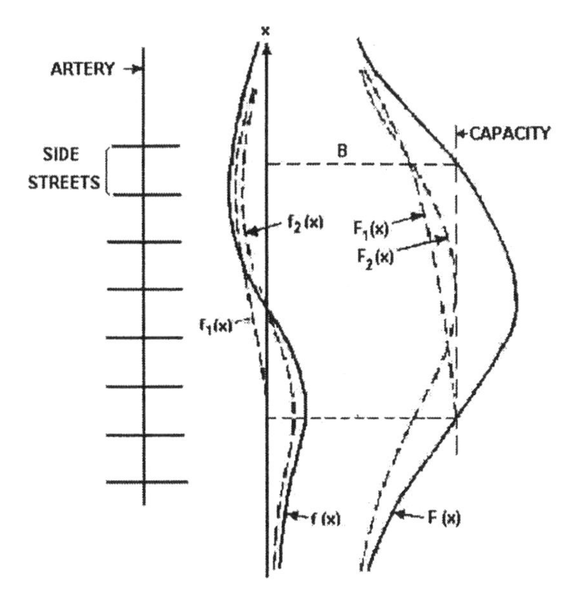

Figure 16. An oversaturated artery

rather than ones with step function changes. If the net input at some intersections, i.e. influx minus outflow of traffic, is positive along an appreciable length of the artery, then the demand along the artery may increase above capacity levels over a section of the artery. In Fig. 16, the demand is shown as a continuous curve $f(x)$. The demand along each point is the integral of this curve, $F(x)$. If

$$F(x) > Q \qquad A < x < B \ , \tag{34}$$

where Q is the maximum throughput of the artery, then congestion will set in along a major portion of the artery. Progression becomes meaningless in this case, but for lack of a better design it is not uncommon in many cities to have the lights synchronized for progression anyway – to the frustration of the traffic engineers and driving public.

Congestion may be avoided by controlling the net input along the artery, and may take one of several forms:

1. A barrel-shaped green through band may be tried, with a bulge in the potentially congested region. Such a band would favor the arterial flow of traffic and also limit the inputs from the side streets. The bulge of the

through band might be made traffic-responsive, changing with traffic demands.
2. One may limit the influx past the point A, making the net input negative past this point, as shown by the input curve $f_1(x)$ and the corresponding demand curve $F_1(x)$ in Fig. 16.
3. One may scale all net inputs down to the curve $f_2(x)$, with a corresponding demand curve $F_2(x)$, so that the maximum of $F_2(x)$ does not exceed the capacity of the artery.

Options 2 and 3 are akin to the "entrance ramp control" which is increasingly applied to freeways to prevent their congestion. All of the above alternatives assume, of course, that there is enough parking space in the side streets, or alternate route choices for the traffic that is blocked from entering the artery. In any case, the optimum solution involving constraining the traffic entering the artery can be obtained on the basis of historical or real time data.

3.5 OVERSATURATED SYSTEMS

In the preceding sections, we considered intersections, and systems of intersections, in which traffic demands are not so high that they cause excessive queueing at any point. However, every major city of today, and even some minor ones, suffer from the experience of "oversaturation" of certain intersections, i.e. the inability of these intersections to serve out all the queues during a green phase, resulting in ever increasing sizes of queues. Here, when we talk of queues, we understand that their position may not be limited a single road segment, but may extend over several of them. In what follows, we discuss systems which become oversaturated during a "rush period", but eventually serve out the built-up queues after the end of the rush period. We begin with the simplest case of a single, relatively isolated, intersection.

3.5.1 A Single Oversaturated Intersection

Let us consider an isolated intersection serving n competing traffic demands. Vehicles arrive at rates $q_j(t)$, $(j = 1,2,\ldots,n)$, and some or all of them are discharged during the green phase favoring their movement. It is assumed that the discharge is at the maximum service rate, or "saturation flow", s_j. When the demand is low, the queues developed during the red phase are completely served during the green phase, near the end of which newly arriving vehicles may go through the intersection without stopping.

The average green time required for serving all the cars arriving during a light cycle c is

$$\bar{g}_j = \frac{\bar{q}_j c}{s_j} , \tag{35}$$

where $\bar{q}_j$ is the average arrival rate of vehicles during the light cycle. As discussed in the preceding section, the standard practice accepted by traffic engineers and managers has been to divide the "total available green time" in proportion to the $\bar{g}_j$ values of Eq. 35, leading to the same "degree of saturation" along all competing directions. Accordingly, the effective green phases g_j satisfy the relationships

$$\frac{\bar{q}_1}{s_1 g_1} = \frac{\bar{q}_2}{s_2 g_2} = \cdots\cdots = \frac{\bar{q}_n}{s_n g_n} . \tag{36}$$

As was shown in the preceding section, in the case of an undersaturated intersection, the rule of thumb shown in Eq. 36 comes very close to optimizing the operation of the traffic light, in the sense of minimizing delays to users.

Let us now consider, as an example, an intersection involving two competing streams, and assume that the demand at the intersection increases until, after some point, one or both of the queues cannot be completely served during the green phases. This happens when the q_j increase, so that

$$\frac{q_1}{s_1} + \frac{q_2}{s_2} > 1 - \frac{L}{c} , \tag{37}$$

where L is the lost time for acceleration and clearance. Equation 37 defines the case of oversaturation, discussed by Gazis and Potts (1965). The allocation of the green may still be done in this case according to Eq. 36, but it is meaningless and very unlikely to minimize delays to the users. In this case, queues build up during the rush period, and are only served completely after the end of this period, provided that the end of one rush period does not merge with the beginning of the next one! Optimization of the operation of the light at an oversaturated intersection must then involve, not the minimization of the delay per cycle, but the minimization of the "aggregate delay to both streams, during the entire rush period.

The situation is shown in Fig. 17, where the cumulative arrival, Q, and cumulative service, G, are plotted versus time, for one of the competing streams. The quantities Q and G are the integrals over time of the arrival rate and service rate, respectively, namely

$$Q(t) = \int_0^t q(\tau)d\tau, \qquad G(t) = \int_0^t \gamma(\tau)d\tau, \tag{38}$$

where $\gamma(\tau)$ is the service rate, and the time origin is taken as the onset of oversaturation. The service curve is actually more like the sawtooth curve of Fig. 17, but it will e drawn as a smooth curve during our discussion, for simplicity.

The area between the curves Q and G is a measure of the aggregate delay to the users of the intersection during the rush period. Their vertical distance at any time t is the effective size of the queue at that time, and their horizontal distance at any point Q, such as $(t_2 - t_1)$ at Q_2, is a measure of delay to an individual vehicle arriving at time t_1. The service rates must cannot be altogether arbitrary, but they must satisfy certain constraints, which depend on the fact that the green phase for each direction must lie between a minimum and a maximum. If g_{min} and g_{max} are the bounds of the green phases, then the service rates γ_j satisfy the constraints

$$g_{\min} \leq g_j = \frac{c\gamma_j}{s_j} \leq g_{\max} \quad . \tag{39}$$

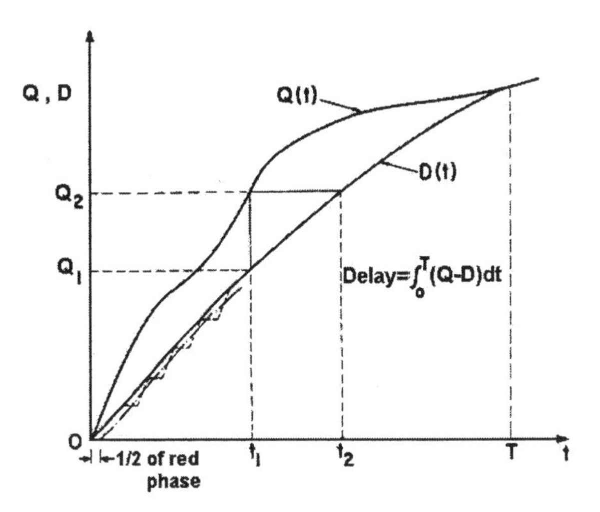

Figure 17. Definition of the basic variables for an oversaturated Intersection.

The optimization of the operation of the traffic light may now be formulated as an "optimal control" problem, with the split of the green phase among the two directions being the control variables. On the basis ofEq. 39, this split may vary between an upper and a lower bound, thus defining a "control region". Any acceptable service strategy must involve time-dependent $\gamma_j(t)$, which satisfy Eq. 39 plus the equation

$$\frac{\gamma_1(t)}{s_1}+\frac{\gamma_2(t)}{s_2}\leq 1-\frac{L}{c} \quad , \tag{40}$$

where the equality applies when sufficiently long queues exist in both directions to supply saturation flow during the entire length of the green phases.

The objective function of the control problem is the aggregate delay to the users, and the statement of the problem is the following:

Minimize the delay function

$$D = \sum_{j=1}^{2} \int_0^T \left[Q_j(t) - G_j(t)\right] dt \ , \tag{41}$$

subject to the conditions in Eqs. 39 and 40 for the values of γ_j, which determine the G_j. The time T is defined by the equation

$$G_j(T) = Q_j(T) \ , \tag{42}$$

and it is assumed that after time T the capacity of the intersection exceeds the demand in both directions.

The solution to the control problem, following Gazis (1964), proceeds as follows, (see Fig. 18):

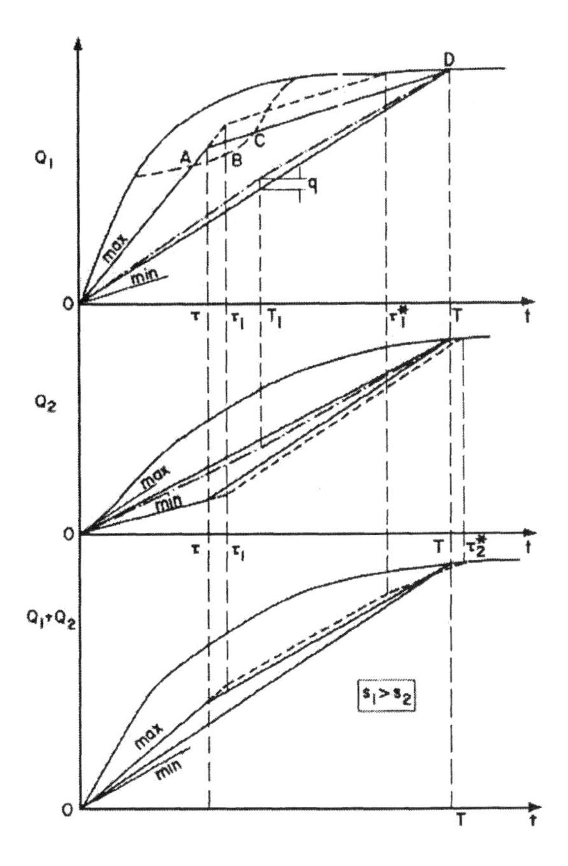

Figure 18. A graphical derivation of the optimum control of an oversaturated intersection

There exists a time T that is the earliest possible end of oversaturation. It can be determined by assuming a single setting for the traffic light during the entire period $0 \leq t \leq T$, which corresponds to constant values of γ_j, such that both queues are exhausted together at time T. Denoting these constant γ_j as γ_j^*, we find that they can be determined, together with T, from the system of equations

$$\frac{\gamma_1^*}{s_1} + \frac{\gamma_2^*}{s_2} = 1 - \frac{L}{c} \ . \qquad (43)$$

$$\gamma_j^* = Q_j(T), \qquad j = 1,2 \ .$$

In Eq. 43, and Fig. 16, the $Q_j(t)$ are known functions of t, based on historical data of rush period traffic inputs for the intersection. The single setting strategy is not the only strategy which exhausts both queues at the same time, neither does it constitute the control which minimizes the aggregate delay. It can be seen in Fig. 18 that we can construct multistage service curves, which also exhaust both queues at the same time, by increasing the delay in one direction and decreasing it in the other. The dash-dot lines in Fig. 18 show such a strategy, in which we give a little more green to direction 1 up to time T_1, and take some away from T_1 until T, so that queue 1 is exhausted at time T. This decreases the delay to stream one, at the expense of an increase in delay for stream 2, but it can be proved easily that queue 2 will also be exhausted at time T.

Let us assume that at time T_1 the queue 1 is decreases, with respect to the single-setting strategy, by the amount q. Queue 2 will be increased at the same time by the amount $(s_2 / s_1)q$. Assuming, without loss of generality, that $s_1 > s_2$, we find that we have reduced the aggregate delay by the amount

$$\delta D = \frac{1}{2}\left(1 - \frac{s_2}{s_1}\right)qT \quad . \tag{44}$$

We can continue trading off delay, in a way that favors the stream with the greater saturation flow, until we reach the optimum strategy, which is a two-stage operation. During the first stage, stream 1 receives the maximum green phase, and stream 2 receives the minimum one, and during the second stage the allocation of the green is reversed. The switch-over point is given by

$$\tau = \frac{(c / s_1)Q_1(T) - g_{\min}T}{g_{\max} - g_{\min}} , \tag{45}$$

with T and $Q_1(T)$ satisfying Eqs. 43. Incidentally, the switching from one extreme of the control variable to the other is generally referred to as *bang-bang control* , a term somewhat less than reassuring when used in connection with traffic.

It can be seen from the preceding discussion that the exact shape of the cumulative demand curves $Q_j(t)$ during most of the rush period does not change the solution, or the minimization result, as long as the cumulative demand curves do not dip below the cumulative service curves, as shown in the top curve of Fig. 18. If the cumulative demand curves can be approximated reasonably well by some asymptotic straight lines near the end of the rush period, we can obtain simple expressions for T ,and for the switch-over time τ. For example, if

$$Q_j(t) = A_j + B_j t \ , \tag{46}$$

then

$$\begin{aligned} T &= \frac{A_1 s_2 + A_2 s_1}{s_1 s_2 (1 - L/c) - (B_1 s_2 + B_2 s_1)} \ , \\ \tau &= \frac{(A_1 b - A_2 a) + u_{\min}(\lambda A_1 + A_2)}{(\lambda a + b)(u_{\max} - u_{\min})} \ , \end{aligned} \tag{47}$$

where

$$\begin{aligned} a &= B_1 \ , \\ b &= B_2 - s_2\left(1 - \frac{L}{c}\right) \ , \\ \lambda &= \frac{s_2}{s_1} \ , \\ (u_{\min}; u_{\max}) &= \frac{s_1}{c}(g_{\min}; g_{\max}) \ . \end{aligned} \tag{48}$$

The total delay to the users may be further reduced, if we prolong a little bit the agony of the drivers in stream 2, with $s_2 < s_1$. The ultimately optimum setting is obtained when the durations of the last straight line portions of the two service curves, satisfy the relationships, (Fig.18),

$$\frac{\tau_1^* - \tau_1}{\tau_2^* - \tau_2} = \frac{s_2}{s_1} \quad . \tag{49}$$

The above relationships regarding the optimal control of the intersection may be obtained by graphical differentiation, i.e. by proving that a very small change, positive or negative, in the time length $\tau_2^* - \tau_1$, together with the associated small change in $\tau_1^* - \tau_1$, produces zero gain in reducing the total delay.

Finally, we address the problem of finding the optimum light cycle, c. The light cycle affects the total throughput of the intersection, because of the lost time L in each cycle, which may be assumed to be constant for any reasonable range of c. The ratio L/c decreases with increasing c, thus increasing the total throughput, and reducing the value of T. However, the additional delay due to intermittent service, i.e. the saw-tooth area in Fig. 17, increases with increasing c. The problem is to find the value of c corresponding to a stationary value for the total delay. This can generally be done for given functions $Q_j(t)$, and a given sequence of settings for the traffic light. Figure 19 shown this derivation for a single setting of the traffic light, and cumulative demand curves approximated by straight lines near the end of the rush period, namely,

$$Q_j(t) \approx \alpha_j + \beta_j t \quad . \tag{50}$$

The appropriate single setting corresponds to green phases, g_j, and service rates γ_j satisfying the relationships

$$\begin{aligned} \gamma_j &= s_j g_j / c \ , \\ \gamma_j T &= \alpha_j + \beta_j T \ . \end{aligned} \tag{51}$$

Now, let

$$g_1 = \lambda c \; ,$$
$$g_2 = (1-\lambda)c - L \; . \qquad (52)$$

Solving for λ and T, we obtain

$$\lambda = \left\{\left[\left(1-\frac{L}{c}\right)s_2 - \beta_2\right]\alpha_1 + \alpha_2\beta_1\right\}(\alpha_1 s_2 + \alpha_2 s_1)^{-1} \; ,$$
$$T = \frac{\alpha_1}{(s_1\lambda - \beta_1)} \; . \qquad (53)$$

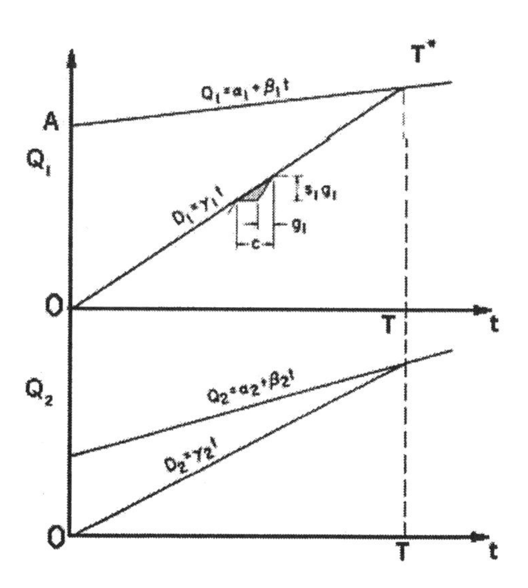

Figure 19. Optimum light cycle, for a single setting during the entire rush period.

The saw-tooth area, per cycle, is equal to

$$d_1 = (c-g_1)\frac{s_1 g_1}{2} + (c-g_2)\frac{s_2 g_2}{2} \; , \qquad (54)$$

hence the total additional delay from 0 to T is

$$D_1 = \frac{Tc}{2}\left\{(1-\lambda)\lambda s_1 + \left(\lambda + \frac{L}{c}\right)\left[1 - \left(\lambda + \frac{L}{c}\right)\right]s_2\right\} . \qquad (55)$$

The total delay over the entire rush period is

$$D = \int_0^T [Q_1(t) + Q_2(t) - (\gamma_1 + \gamma_2)t]dt + D_1 \ , \qquad (56)$$

or

$$D = \int_0^T [\alpha_1 + \alpha_2 + (\beta_1 + \beta_2 - \gamma_1 - \gamma_2)t]dt + D_1 - D_0 \ , \qquad (57)$$

where D_0 is the (constant) difference between the trapezoidal, cumulative demand curve, such as OAT^* of Fig. 19, and the actual demand curve, which is not plotted in that figure. Finally,

$$D = (\alpha_1 + \alpha_2)T + (\beta_1 + \beta_2 - \gamma_1 - \gamma_2)\frac{T^2}{2} + D_1 - D_0 \ . \qquad (58)$$

The optimum cycle is obtained by differentiation. i.e.,

$$\frac{dD}{dc} = [(\alpha_1 + \alpha_2) + (\beta_1 + \beta_2 - \gamma_1 - \gamma_2)T]\frac{dT}{dc} - \frac{T^2}{2}\frac{d(\gamma_1 + \gamma_2)}{dc} + \frac{dD_1}{dc} = 0 \ . \qquad (59)$$

The general case involves a fair amount of algebraic manipulation. Here, we give as an example the derivations for the simple case when

$$\beta_1 = \beta_2 = 0 \ , \qquad (60)$$

in which case, the optimum cycle turns out to be

$$c_0 = L + \left(\frac{\alpha_1}{s_1} + \frac{\alpha_2}{s_2}\right)\left[L\frac{(\alpha_1+\alpha_2)s_1 s_2}{(s_1+s_2)\alpha_1\alpha_2}\right]^{1/2} . \tag{61}$$

For example, if $\alpha_j / s_j = 40$ min., and $L = 3$ sec., we obtain

$\lambda = 0.5$, $T = 80$ min., and $c_0 = 2.3$ min.

Of course, such a choice for c is acceptable only if the range of the cycle includes the derived optimum cycle of 2.3 min.

Let us briefly discuss the meaning of the above derivation of optimal control of the oversaturated intersection. Our derivation shows that, during the early part of the rush period, the direction with the higher saturation should be favored to the highest possible degree. At some point in the midst of the rush period, service should switch to maximum for the direction with the lower saturation flow. If the timing is selected properly, both queues will be served out at the same time, and with the minimum aggregate delay to the users. Incidentally, this solution is far from obvious, or intuitive. This author was told by a British traffic expert that "in England we actually favor the poor devil on the side street who competes for service against a big stream of traffic – sort of, rooting for the underdog". It turns out that such compassion, or even a democratization of traffic control equalizing the rate of discharge of queues, fails to optimize the system. The reason can be seen from the schematic diagrams of Fig. 20. The intersection, viewed as a system, receives a traffic demand rate higher that the attainable service rate during the rush period. This means that the cumulative service curve lies

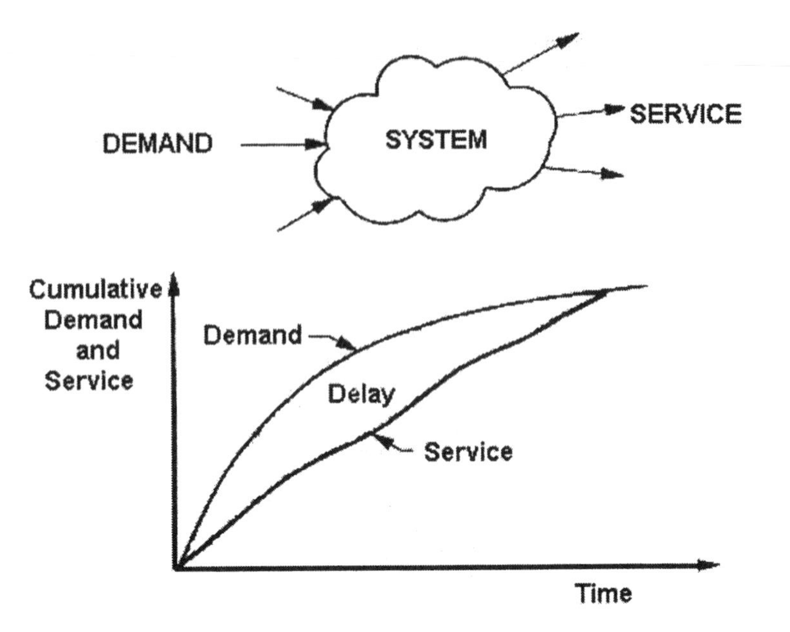

Figure 20. The origin of congestion at an intersection

below the cumulative demand curve during the rush period, ad the area between these two curves is a measure of the aggregate delay to the users. The best way of reducing this aggregate delay is by maximizing the service rate as much as possible during the early pert of the rush period, with proper adjustments during the last part of the rush period for the purpose of ending congestion simultaneously in both directions. The operative words here are that we must *get those cars out of here as soon as possible*. It will be seen later on in this chapter that this principle applied to even more complex oversaturated systems.

A more rigorous mathematical derivation

So far, we have limited our approach to manipulating the details of a graph, in effect performing graphical differentiations and the like. A more complete solution, given by Gazis (1964), is cast in the framework of Control Theory, and uses the Pontryagin "maximum principle", (see Pontryagin et al, 1962).

Let us refer to Fig. 21, which shows the cumulative demand and service curves versus time, for two competing demands. The queue sizes x_1 and x_2

are the *state variables* of the system, and they satisfy the differential equations

$$\frac{dx_1}{dt} = q_1(t) - u \ , \qquad (62)$$
$$\frac{dx_2}{dt} = q_2(t) - \frac{s_2}{c}(1 - L) + \frac{s_2}{s_1}u \ ,$$

where u is a *control variable* defined by

$$u = \frac{s_1 g_1}{c} \ . \qquad (63)$$

We seek to minimize the aggregate delay to the users, given by

$$x_0 = \int_0^T (x_1 + x_2)\,dt \ , \qquad (64)$$

subject to the end conditions

$$x_j(0) = x_j(T) = 0 \qquad j = 1,2 \ , \qquad (65)$$

and the constraint that the control variable u be within the admissible control domain, i.e.

$$\frac{s_1 g_{\min}}{c} = u_{\min} \le u \le u_{\max} = \frac{s_1 g_{\max}}{c} \ . \qquad (66)$$

This is a problem in linear control theory, and the linearity leads to a bang-bang solution. Following Pontryagin, we see that Equations 62 and 64 are equivalent to

$$\frac{dx_i}{dt} = f_i(x_j, u) \quad \begin{matrix} i = 0,1,2 \\ j = 1,2 \end{matrix} \quad , \tag{67}$$

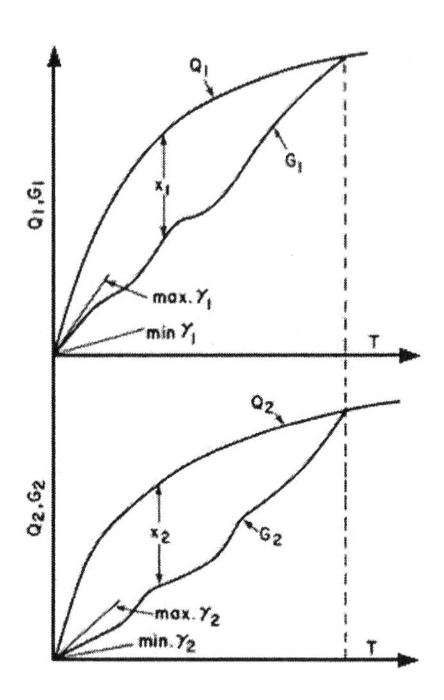

Figure 21. Derivation of optimal control of an oversaturated intersection, using the Pontryagin maximum principle

where

$$\begin{aligned} f_0 &= x_1 + x_2 \ , \\ f_1 &= q_1(t) - u \ , \\ f_2 &= q_2(t) - s_2\left(1 - \frac{L}{c}\right) + \frac{s_2}{s_1}u \ . \end{aligned} \tag{68}$$

We now define the *adjoint variables* ψ_i, satisfying the relationships

$$\frac{d\psi_i}{dt} = -\sum_{l=0}^{2} \frac{\partial f_l(x_i, u)}{\partial x_i} \psi_i \qquad i = 0,1,2 \quad . \tag{69}$$

The system of Eqs. 68 can be solved, yielding

$$\begin{aligned} {}_0 &= k_0 \quad , \\ {}_i &= -k_0 t + k_i \qquad i = 1,2 \quad , \end{aligned} \tag{70}$$

where k_0, k_1, and k_2 are constants. We now define the Hamiltonian of the system, H, given by

$$H(\psi_i, x_i\, u) = 2\sum_{l=0} \psi_i f_i = zu + w \quad , \tag{71}$$

where

$$\begin{aligned} z &= \left(1 - \frac{s_2}{s_1}\right) k_0 t + \left(-k_1 + \frac{s_2}{s_1} k_2\right) \quad , \\ w &= q_1(-k_0 t + k_1) + \left[q_2 - s_2\left(1 - \frac{L}{c}\right)\right](-k_0 t + k_2) + k_0(x_1 + x_2) \quad . \end{aligned} \tag{72}$$

According to Pontryagin's maximum principle, the optimum control maximizes the Hamiltonian, subject to appropriate end conditions. From Eq. 71, we see that the Hamiltonian H is maximized if u is equal to its upper bound when z is positive, and equal to its lower bound when z is negative. Using the *signum* function of z, (sg z = sign of z), we can write the optimum u in the form

$$u = \frac{1}{2}\left[(1 + \mathrm{sg}\, z\,)u_{\max} + (1 - \mathrm{sg}\ z)u_{\min}\right] \quad . \tag{73}$$

The remainder of the computations toward the optimum solution are straightforward, leading to the expressions already given for the switch-over time τ and the duration of the rush period, T.

3.5.2 Complex Oversaturated Systems

The preceding discussion of the optimal control of a single oversaturated intersection makes sense if this intersection is essentially alone in causing congestion, and the rest of the traffic network is essentially undersaturated. It could also be valid if other oversaturated points in the network are relatively far from the intersection in question, thus being *uncoupled* from it. Now, let us consider some systems, some involving neighboring intersections, and seek their optimal control.

a) A Pair of Intersections

When two oversaturated intersections are close together, they are coupled through the number of vehicles that pass through both intersections in sequence. Gazis[42] considered the system of two intersections serving the traffic streams 1, 2, and 3, where the stream 1 is assumed to go through the pair of intersections (1, 2) and (1, 3) without turning, (Fig. 22). Let us assume that the saturation flows in the three directions satisfy the relationships

$$s_1 > s_2 \,, \qquad s_1 > s_3 \,, \qquad s_1 < s_2 + s_3 \quad . \tag{74}$$

If we optimize the two intersections independently, following the preceding discussion of the optimization of the single intersection, we will conclude that stream 1 must receive preferential treatment in both intersections. However, if we take the point of view that streams 2 and 3 are like traffic streams in a single traffic stream moving along a *boulevard*, then stream (2 + 3) should receive the preferential treatment, on the basis of the last of Eqs. 73. It turns out that this second point of view is the correct one, because in the first one we essentially count the delay to stream 1 twice, once at intersection (1, 2), and then again at intersection (2, 3). The correct solution proceeds as follows:

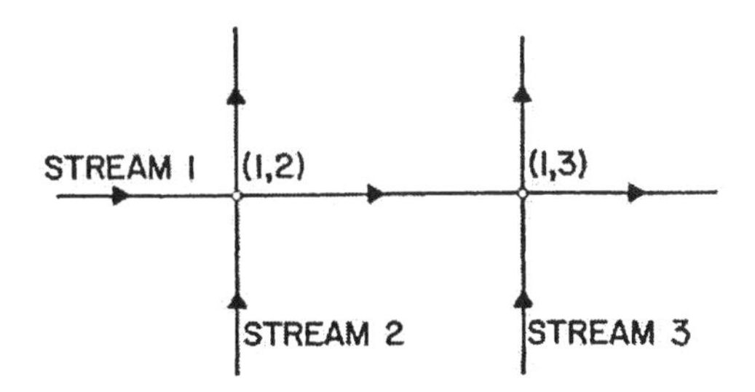

Figure 22. A pair of oversaturated intersections

Let us assume that the transit time and queueing storage between the two intersections are negligible relative to the other retained variables of comparable nature, and that the two intersections become oversaturated roughly at the same time. We can again obtain an optimal policy of allocation of green phases by applying the same policy improvement scheme used for the single intersection, namely, trading off delays of the various streams if they reduce the aggregate delay. First, we obtain the earliest end of the rush period for each one of the two intersections, viewed as isolated. Let these ends be the times T_{12} and T_{13} for intersections (1, 2) and (1, 3), respectively. We also compute the switch-over times corresponding to optimizing the operation of these two intersections as isolated ones. These switch-over times are given by

$$\tau_{1j} = \left[g_{\max} T_{1j} - \left(\frac{c}{s_1} \right) Q_1 T_{1j} \right] (g_{\max} - g_{\min})^{-1} \qquad j = 2,3 . \tag{75}$$

If the maximum and minimum service rates for stream 1 are the same at both intersections, the relative magnitude of τ_{1j} is the same as that of T_{1j}. We now obtain the complete solution to the problem, shown in Fig 23, as follows:

Since streams 2 plus 3 must receive preferential treatment at the start of the rush period, we find the green allocation favoring these streams, for each of the intersections treated as isolated ones. The corresponding service for intersection (1, 2) is the optimum, because any delay trade-off reducing the delay for stream 1 increases the delay to the combined streams 2 and 3 by q greater amount. However, after the switch-over time τ_{12}, a delay trade-off is possible between streams 1 and 3. The maximum trade-off corresponds to a reduction of the delay to stream 1 by the area of the parallelogram (ABCD), with a corresponding increase of the delay for stream 3 by the area (abcd), which is smaller than (ABCD) since we have assumed that $s_1 > s_3$. The key to the optimal solution is that we can favor stream 1 when it competes with only one of the streams 2 and 3, but we must favor the combination of streams 2 and 3 if they compete with stream 1 simultaneously.

As a consequence, the optimum control of the pair of intersections involves, in general, three switch-over times, one at intersection (1, 2), and two at intersection (1, 3). It is tacitly assumed that there is enough parking space along the direction 1 between the two intersections to accommodate the queue before intersection 3. Otherwise, yet another adjustment to the strategy is called for, which will not be discussed here. A complete discussion of various aspects of the problem is given by Gazis (1964).

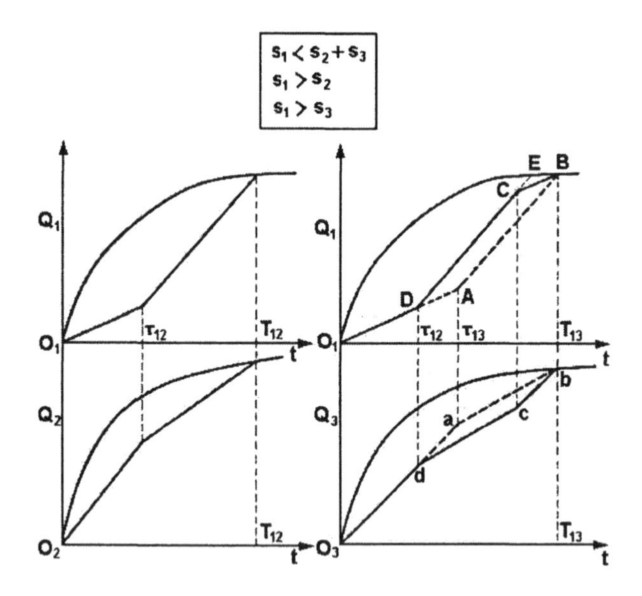

Figure 23. A pair of two neighboring oversaturated intersections

The preceding discussion in Sections 3.5.1 and 3.5.2 defines the operation of network elements (intersections), and entire networks of intersections, which are *Store-and-Forward Networks*. In such networks, during the rush period, traffic queues up at some or all of the intersections, which can be described as *Throughput Limiting Points (TLPs)*. Optimizing the operation of the traffic lights of the network comprises a selection of a sequence of green phase allocation in each intersection, which minimizes the aggregate delay to the users. The examples given in the preceding section show that the optimal strategy involves transitions from one apex of the control domain to another. We will now consider some additional systems.

b) Three Intersections Forming a Triangle

The optimal control of a system of intersections forming a triangle was also considered by Gazis (1964) and shown in Fig. 24. Let us assume that traffic moves along the directions 1 to 3 without turning movements, and the cumulative demands are given by

$$\begin{aligned} Q_1 &= 1800 + 500t \ , \\ Q_2 &= 1440 + 300t, \\ Q_3 &= 1300 + 300t \ . \end{aligned} \tag{76}$$

where t is the time, given in hours. Let us assume that the saturation flows are $s_1 = 3600$, $s_2 = 2400$, and $s_3 = 2000$ vehicles per hour. We also assume that the maximum green phase is 60 s and the minimum 20 s, with 10 s allowed per cycle for clearing the intersection at phase changes.

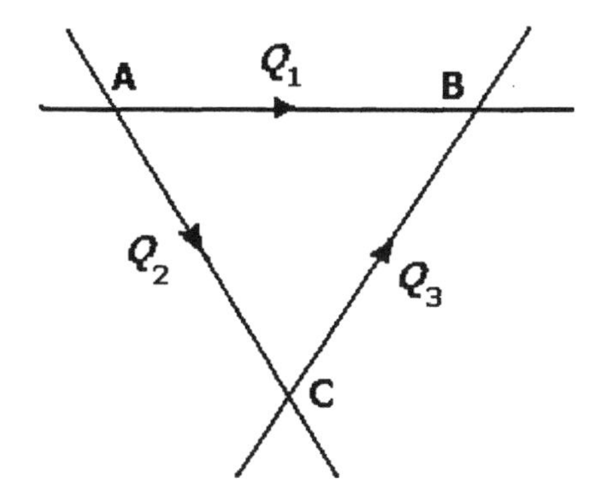

Figure 24. A system of three interacting, oversaturated intersections

The admissible control domain is found to be one bounded by the hexahedral surface OABCD shown in Fig. 25.

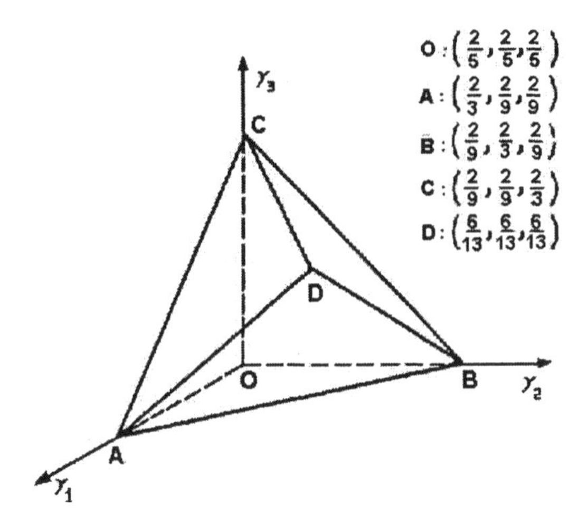

Figure 25. Admissible control domain for the system shown in Fig.24

In Fig. 25, the variables γ_i are the fractions of the light cycle allocated to the three traffic streams. It is assumed that a common cycle is used at all intersections, hence

$$\gamma_i = \frac{g_i}{c - L}, \tag{77}$$

where g_i is the usable green phase for the ith stream, and L is the 10 s time allowed for clearance at phase change. For example, the apex A of Fig. 25 corresponds to $g_1 = 60$ s and $g_1 = g_1 = 20$ s. The usable green phase for streams 2 and 3 is applied where these two streams intersect stream 1, and they produce streams which do not constitute congestion at their intersection, since the sum of the two greens is smaller that the available green at that intersection. Thus, the specific allocation of the green at that intersection has no influence on the performance of the intersection.

It may be easily ascertained that during the rush period at least one of the green phases must be equal to the maximum, in order to decrease the effect of the lost time for clearance. This means that only the apexes A, B, C, and D need be considered in seeking an optimum solution. Table 1 shows the service rates for the three streams, for each choice among the four apexes, listed in order of decreasing total throughput of the system.

Table 1 . Service rates corresponding to the four apexes of the control domain shown in Fig. 25.

	Apex	Stream 1	Stream 2	Stream 3	Total
D	green phase	60	60	60	$c = 130$
	service rate	1660	1110	920	3690
A	green phase	60	20	20	$c = 90$
	service rate	2400	530	440	2970
B	green phase	20	60	20	$c = 90$
	service rate	800	1600	440	2840
C	green phase	20	20	60	$c = 90$
	service rate	800	530	1330	2660

The traffic light settings corresponding to the four apexes are assumed to be maintained for times T_A, T_B, T_C , and T_D , and it is required that all three queues be served out simultaneously at time $T = T_A + T_B + T_C + T_D$.

We have seen from the preceding examples that the optimum control policy must produce a concave curve for the cumulative service. The same principle applies here, and so we are led to a procedure for finding the optimum policy, which can be formulated as a linear programming (LP) problem, the following:

Minimize

$$T = T_A + T_B + T_C + T_D \ , \tag{78}$$

subject to

$$M_{ij}x_j = b_i + c_iT \ , \tag{79}$$

where

$$M = \begin{bmatrix} 2400 & 800 & 800 & 1660 \\ 530 & 1600 & 530 & 1110 \\ 440 & 440 & 1330 & 920 \end{bmatrix} ,$$
$$b = (1800,\ 1400,\ 1300) , \tag{80}$$
$$c = (500,\ 300,\ 300) ,$$

and the unknowns x_i must be non-negative. The solution to this LP problem, for the specific values given above, is

$$T_A = 0$$
$$T_B = 0.124\,hr$$
$$T_C = 0.394\,hr$$
$$T_D = 1.43\,hr \ .$$

Furthermore, to ensure concavity of the cumulative service curve, the sequence of these time segments must be in decreasing order of the corresponding service rates, namely

$$T_D \rightarrow T_B \rightarrow T_C \ .$$

A key observation should be made at this point. At the beginning of the rush period, favoring any one of the streams, usually the one associated with the highest saturation flow **does not** optimize the operation of the system, because it “starves” the downstream intersection of the unfavored streams. It can be seen from the above solution that the best policy starts with equal splits for all streams. Yet, as of the writing of this book, if we drive around many congested areas around the world, we are likely to find that the local traffic authorities set traffic lights at such triangle systems in proportion to the saturation flows, being unaware of the consequence of such a policy in the case of oversaturated systems.

It may be noted that the LP formulation can be used without knowledge of the exact form of the admissible control domain, and hence the possible number of distinct control periods, and corresponding service rates. The time axis may simply be divided into equal segments, and the resulting queue sizes on various links can be used as the unknowns of the LP problem, with the objective of minimizing aggregate delay to the users of the system. The solution to this problem is a suboptimal one, approaching the optimal solution as the size of the time segments is decreased. This approach will be used later on in this section for the optimization of a complex transportation network.

c) Spill-back from an Exit Ramp of an Expressway

Let us consider the system shown in Fig. 26, comprising an expressway 1, an exit ramp 2, and the exit highway. The exit highway may also be an expressway, in which case the exit ramp may lead into an acceleration lane. Here, we are interested in exploring a situation in which the rate of discharge from the exit ramp into the exit highway is controlled. We consider the case discussed by Gazis (1965), in which the exit highway 3 is a major artery but not an expressway, and the intersection of 2 and 3 is controlled by a traffic light, a very common situation in many urban areas. If 3 is an expressway, and input into it is controlled through "entrance ramp control", the discussion that follows is applicable without major changes.

If the intersection of 2 and 3 becomes oversaturated, then a queue builds up along the exit ramp. Because of the limited parking capacity of this ramp, the queue spills into the expressway, often with devastating results. Not only does the spilled queue block one of the expressway lanes, but often, drivers from the neighboring lane attempt to enter the exiting stream, slowing down and obstructing the movement of the through traffic. It is not unusual to have a three-lane expressway degenerate into barely a one-lane roadway, causing inordinately large tie-ups upstream of the exit ramp. Cases such as this one abound in many metropolitan areas as of the time of the writing of this book. The question is, can one minimize the impact of this phenomenon by properly operating the traffic light 4 , in a way that minimizes the delay to the combined streams 1, 2, and 3?

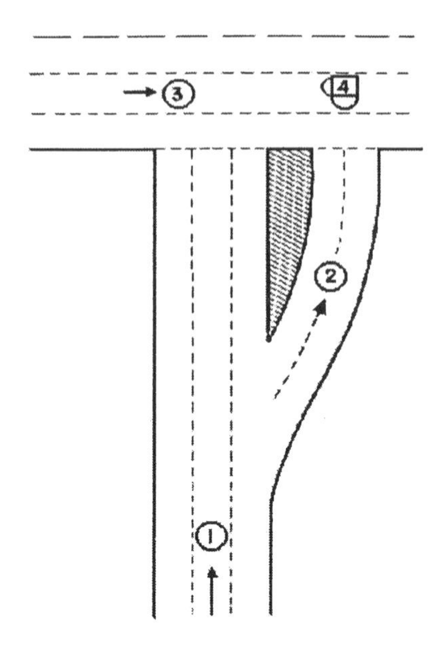

Figure 26. Spill-back from an exit ramp of an expressway

The cumulative demand curves Q_1, Q_2, and Q_3, as well as the maximum and minimum service rates for all streams, are shown in Fig. 27. The service rates for streams 2 and 3 are determined directly from the split of the traffic light, with Γ_j being the maximum, and γ_j the minimum service rate. The expressway is assumed to have a throughput equal to Γ_1 when unobstructed, and γ_1 when clogged due to spill-back. More often than not, the saturation flows s_2 and s_3 are such that

$$s_2 < s_3 \ . \tag{81}$$

Therefore, according to the discussion given earlier concerning an isolated, oversaturated intersection, the optimum operation of the trafficlight 4 would involve service curves such as O_3EF and O_2ef, with the highway stream 3 receiving preferential treatment. However, in the present case the ramp stream 2 is coupled with the stream 1 along the expressway through the spillback phenomenon. If the storage capacity of the ramp is Q_2^*, then spillback will take place from time t_g to time t_h, where these times are determined by drawing the curve

$$\overline{Q}_2 = Q_2 - Q_2^* \,, \tag{82}$$

and finding the intersection of this curve with O_2ef.

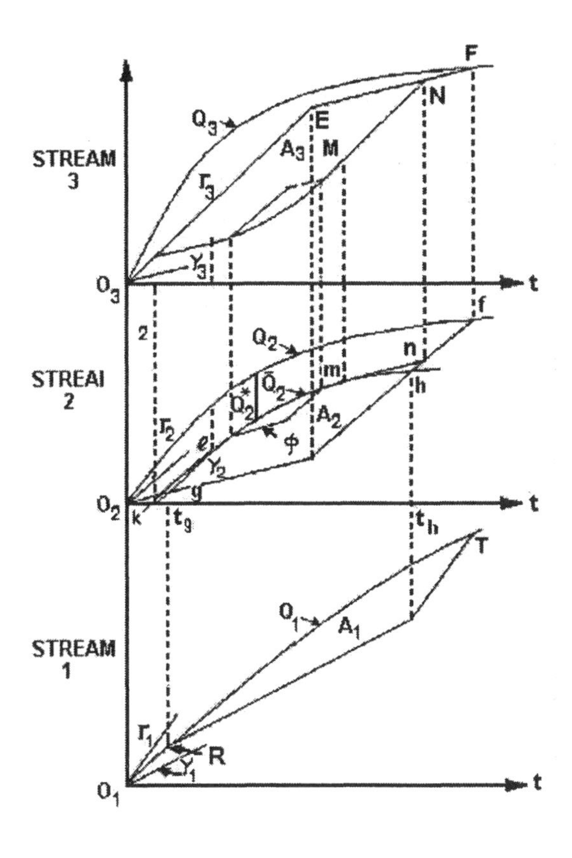

Figure 27. Optimum control of the system comprising streams 1, 2, and 3.

We can see from Fig. 27 that if we choose the service curves O_2klmnf and O_3KLMNF, we can eliminate spillback altogether. The delay to stream 3 is increased by a quantity A_3, the delay to stream 2 is decreased by A_2, and the delay A_1 which would have been caused by spillback is completely eliminated. It should be pointed out here that such a situation is not always possible in many situations of this type, largely because of a very limited capacity Q_2^*. Let us continue with the assumption that it is possible. Since

$$\frac{A_2}{A_3} = \frac{s_2}{s_3} \quad , \tag{83}$$

the total change in the aggregate delay will be

$$\delta = A_1 + A_2 - A_3 = A_1 + A_2\left(1 - \frac{s_3}{s_2}\right) \quad . \tag{84}$$

If δ is positive, we have a net reduction of the aggregate delay, and therefore it pays to adopt a strategy of keeping the queue along the ramp below the critical value Q_2^*, if at all possible. If δ is negative, then spillback is not as damaging as it appears, at least in terms of total delay, which is minimized by an optimum operation of the light 4, assumed isolated. However, it may still be desirable to prevent, or at least limit the congestion on the expressway for safety reasons, which may override delay considerations. If this is the case, one may accept a small negative delay trade-off δ.

If the delay criterion is the dominant one, we find that a critical constant rate of demand q_1 along the expressway exists, which is related to A_2 according to a relationship obtained by setting δ in Eq. 84 equal to zero, leading to the relationship

$$A_1 = \frac{(\Gamma_1 - \gamma_1)(q_1 - \gamma_1)}{\Gamma_1 - q_1}\frac{\tau^2}{2} = A_2\left(\frac{s_3}{s_2} - 1\right) \quad , \tag{85}$$

where

$$\tau = t_h - t_g \quad . \tag{86}$$

Solving Eq. 85 for q_1, we obtain

$$q_1 = \frac{2A_2(s_3/s_2 - 1)\Gamma_1 + \gamma_1(\Gamma_1 - \gamma_1)\tau^2}{2A_2(s_3/s_2 - 1) + (\Gamma_1 - \gamma_1)\tau^2} \quad . \tag{87}$$

If the demand rate is smaller than q_1, then spillback is the lesser of two evils, since it corresponds to minimum aggregate delay. By similar arguments we may investigate the possibility of allowing spillback during a portion of the interval $(t_h - t_g)$. If the rate of demand along the expressway falls sufficiently below Γ_1, it may be profitable, in terms of delay, to adopt a strategy such as that shown by the dashed line ϕ in the middle diagram of Fig. 27, and the complementary service curves for streams 1 (not shown in the figure), and 3. Such a policy will introduce some additional delay to the stream 2, and because of spillback it will also delay some vehicles along the expressway, but it will reduce the delays to stream 3. The net change can be computed by an expression similar to Eq. 84, and the policy may be adopted if it is found to be indeed profitable. Finally, if the demand rate along the expressway falls below γ_1, it is always profitable to allow spillback.

A few remarks are in order at this point. One key observation is that intersection 4 should not, in principle, be operated as an isolated one. Many such intersections are operated as isolated, resulting in unduly large delays along the expressway. Indeed, in most cases, the demand along the expressway is so great that it pays to try and eliminate spillback for as long as possible. Another observation is that the storage capacity of the exit ramp is a very crucial quantity. Improvements to the system should probably place a high priority to increasing this capacity; for example, by building a long extra lane leading to the exit ramp, and advising drivers that they can exit only from that lane, thus preventing the blockage of the neighboring lanes by exiting vehicles.

3.6 OVERSATURATED STORE-AND-FORWARD NETWORKS

All the preceding examples of oversaturated systems may be viewed as special cases of an oversaturated store-and-forward network, such as that shown in Fig. 28. In such a network, traffic does not move freely along certain paths, because its movement is obstructed by the presence of

Throughput Limiting Points (*TLPs*) . The most common nature of TLPs is that they are network nodes in which the output capability is lower than the sum of inputs at that node. As a result, queues are formed upstream of a TLP, which are not dissipated until the end of the rush period. In Fig. 28, the nodes C, D, E, F, and G are all TLPs, causing the formation of upstream queues on some or all the arcs converging at those nodes. It may be pointed out that all the oversaturated systems discussed in the preceding sections can be viewed as store-and-forward networks. For example, the single intersection can be viewed as the store-and forward network shown in Fig. 29, with B being the TLP causing the queue formation along the two competing directions.

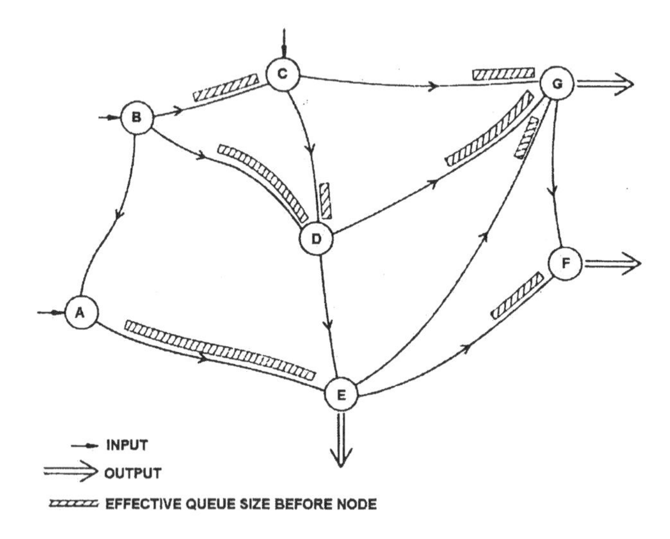

Figure 28. An oversaturated store-and forward network

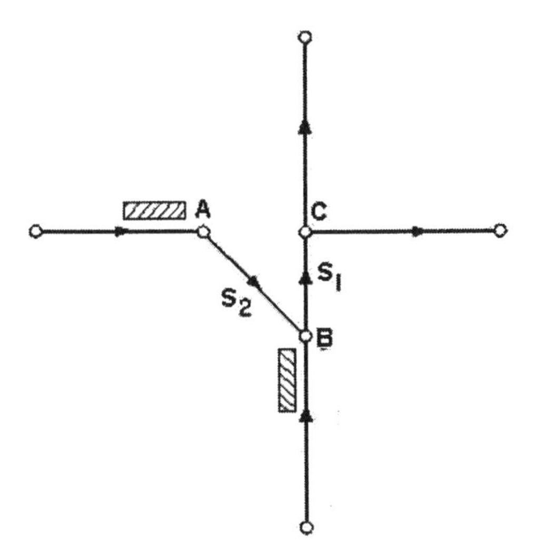

Figure 29. A single, oversaturated intersection viewed as a store-and-forward network.

It should be mentioned at this point that the existence of TLPs is more common than one might think. Many studies of networks involve assumptions of network performance based on phenomenological flow versus concentration relationships. In such studies, it is customary to simply compute an effective speed of traffic as a function of the density along a network arc. The point made here is that, more often than not, this (high) density is the result of the existence of a TLP, causing queueing. As a result, the store-and-forward treatment of an urban network is not just one of many choices, but it is a necessity, since most urban networks become store-and-forward networks, even before heavy congestion sets in.

For the purpose of treating the optimization of the management of a store-and-forward network, we assume the following:

1. There exist some time-dependent, possibly stochastic, origin-destination requirements of traffic volumes, which cannot be fully accommodated during the rush period, because of the existence of TLPs.
2. With each arc of a network are associated three parameters
   - A (fixed) travel time, (or travel "cost"),
   - A capacity restraint, i.e. the maximum possible volume of traffic along the arc per unit of time, and
   - A storage capability allowing traffic to be stored just before a node at the end of an arc, and be dispatched when roadway space becomes available.

3. Traffic moves from node to node at the constant speed corresponding to the arc connecting the nodes, and then either moves past the node, or is stored there.

With respect to item 3 above, it is of course true that queued traffic is stored along the arc before the constricting node. However, for the purpose of optimization, it can be considered as stored in a "bin" right at the node, and then served on a FIFO (First-In-First-Out) basis.

In optimizing the operation of a store-and-forward network, the primary objective is to minimize an aggregate delay function, possibly weighted for priority assignments, subject to operational and capacity constraints. The optimization will certainly comprise finding the optimum *switching* at the nodes, allocating passage through the nodes to various traffic streams. For even better results, *route assignment* for individual vehicles, or platoons of vehicles, could decrease delays even further. To see the possible beneficial effect of route assignment, let us look at Fig. 30, which shows an particularly simple case of a network of one way streets, whose intersections are controlled by traffic lights.

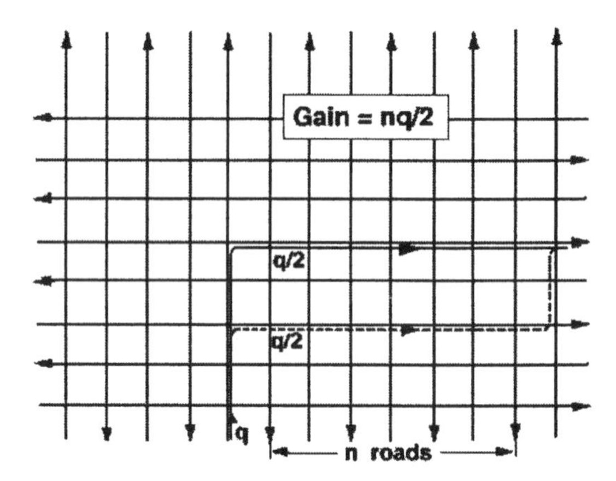

Figure 30. Throughput increase through route assignment

For simplicity, let us assume that all streets are essentially identical and carry about the same amount of traffic, which is pretty close to saturation, therefore the corresponding traffic streams require the entire length of he green phase in order to be served without queueing extending to the next cycle. With equal demands in both directions, the network would operate optimally with equal green phases in all directions.

Let us now assume that at some time a certain portion of northbound traffic along one of the streets, corresponding to a flow rate q wants to turn right and go eastbound at some intersection. If we are to accommodate this turning movement, we must take away a certain portion of the green phase from $(n + 1)$ north/south streams intersected by the right-turning stream. This would result in a loss of overall throughput of the network equal to $(n + 1)\, q$. If, however, we split the stream in two, and merge it again at the $(n+1)^{th}$ street, we gain back half of the lost capacity, or $qn/2$. The author has frequently referred to this strategy as one using "The FLF (Five Loaves, Few Fish) Algorithm". This somewhat irreverent biblical reference to the miracle of feeding a multitude with five loaves and a few fish is used to dramatize the effect of stretching the capability of scarce resources, which in this case is accomplished through route assignment. The point here is that such route assignment increases the overall throughput of the system. If we look at Fig. 20, which may be viewed as a generic figure for all congested systems, we see that increasing the aggregate throughput of the system can only decrease the aggregate delay to the users.

The solution to the complete problem, involving both switching at the nodes and route assignment, is rather cumbersome and will not be given here. Here, we shall assume that the route assignment for all traffic is given, and we shall seek only optimum switching, which regulates the queue discharge rates at various nodes. This switching problem can be cast in the framework of control theory. The state variables are the queues before the nodes, which have originated from upstream nodes including the input nodes, and they must be dispatched optimally toward their destination. The rate of change of these queues depends on the traffic input rates into the network, and on some control variables which represent the switching at the nodes. The objective function is the aggregate delay to the users, which is a function of the state variables. Optimization is achieved if we minimize this objective function, subject to appropriate constraints on the control variables. The mathematical formulation is the following. Minimize

$$D = \sum_j \int_0^T \alpha_j Q_j \, dt \qquad j = 1, 2, \cdots\cdots, J \tag{88}$$

given that

$$\frac{dQ_j}{dt} = f_j(I_k, u_l), \quad (k = 1, 2, \cdots\cdots, K) \;, (j = 1, 2, \cdots\cdots, J) \tag{89}$$

subject to some end conditions corresponding to some given values for Q_j (0) and Q_j (T), and also subject to the constraints

$$u_l \in U \ . \tag{90}$$

In the above equations, Q_j are the state variables, α_j are weighting factors, I_k the (given) traffic inputs, with known destinations and route assignments, u_l the control variables, U the control domain, and f_j some functionals which may possibly involve not only the instantaneous values of $I_k(t)$ and $u_k(t)$ but also their history. If we now discretize the time and replace the time derivatives by differences, and the integral of Eq. 88 by a sum, we are led to a mathematical programming problem, the following.

Minimize

$$D = \sum_{j,n} \alpha_j Q_{j,n} \ , \quad (j = 1,2,\cdots\cdots,J)\,,\ (n = 1,2,\cdots\cdots,N)\ , \tag{91}$$

given that

$$Q_{j,n+1} - Q_{j,n} = f_j\left(I_{k,n}, u_{l,n}\right) \quad \begin{array}{l} k = 1,2,\cdots\cdots,K \\ j = 1,2,\cdots\cdots,J \\ n = 1,2,\cdots\cdots,N \end{array} \tag{92}$$

and subject to the end conditions on $Q_{j,0}$ and $Q_{j,N}$, plus the constraints on the control variables

$$u_{l,n} \in U \ . \tag{93}$$

It is reasonable to assume that the functionals in Eq. (92) are linear in $u_{l,n}$, leading to an Linear Programming (LP) problem of optimization. The optimal solution to this problem minimizes delays, assuming the given route assignments to traffic. One possible approach to obtaining the global optimal solution, including route assignments, is to make small variations to route assignments, trying to bypass the points of maximum congestion, and solve the LP problem repeatedly until we obtain a reasonable approximation to the global optimum.

A number of additional contributions to the control optimization of store-and-forward networks have been made over the years, including those of

D' Ans and Gazis (1976), Singh and Tamura (1974), Michalopoulos and Stephanopoulos (1977a, 1977b), Lim et al (1981), Davison and Ozguner (1983), Park et al (1984), Rathi (1988), Kim and Bell (1992), and Diakaki et al (1999). However, application of the approach has been limited in practice, probably because many theoreticians and practitioners do not realize that **all** traffic networks are of the store-and-forward variety, as soon as even one TLP appears, let alone many of them. Many theoretical contributions continue to use phenomenological relationships to describe flow behind such TLPs, which is at best a rather pointless exercise. Links upstream of a TLP do not miraculously set themselves to accommodate heavy densities by obeying these phenomenological relationships. They simply pile-up traffic into queues, which are served at the discharge rate supplied at the TLP. Modeling of the details of flow upstream of the TLP only imposes computational burdens, without making any contribution to the solution of the problem. Papageorgiou (1999) cites one application of a variant of the store-and-forward approach known as *TUC*, to which he and his colleagues have contributed. The application took place in Glasgow, Scotland, with very successful results.

3.7 FREEWAY CONTROL

Freeways frequently become congested, and even attain a store-and-forward status. Congestion may be caused by something happening along the freeway, it may be the result of excessive inputs from entrance ramps, or it may be the result of spillback from an exit ramp discussed in Section 3.5.2c . An example of something happening along the freeway that may cause congestion is the case of the "Moving and Phantom Bottleneck" discussed in Chapter 1 . Such a bottleneck may be caused by a truck unable to negotiate a steep uphill portion of the freeway, which in this case does not have an extra "slow traffic" lane. A very special example of such an effect will be discussed in the following section, for the case of the Lincoln tunnel of New York City.

Excessive input from an entrance ramp has been addressed by metering the ramp input and restraining it, in order to prevent freeway congestion. Ramp metering may be fixed-time, adjusted for expected traffic conditions at particular times of the day, based on historical data. Obviously, fixed-time ramp metering may miss the mark, since traffic may deviate appreciably from historical data. However, any reduction of input of traffic into the freeway is likely to reduce its congestion, albeit without preventing it completely.

A better approach to ramp metering is through closed-loop control based on real-time measurements of traffic along the freeway. Many such schemes have been suggested and tested, generally showing good results. Papageorgiou (1999) gives a thorough discussion of such schemes, including some advanced methods like the *ALINEA* strategy suggested by Papageorgiou et al (1991).

Generally, all these control strategies do not account for what happens behind the ramp entrance. If the entrance is from surface streets, it is tacitly assumed that it does not matter what happens to them, as long as one gets some improvements for the freeway operation. This is not the case, however, if the ramp input comes from another freeway. In Section 3.5.2-c, we discussed the disastrous effects on the supplying freeway, which an obstruction of the ramp outflow such as ramp metering can cause. Clearly, the best solution for ramp metering should consider the entire system, and not an isolated point of the freeway. Here are some examples of what can go wrong with consideration of isolated input ramps:

- The freeway may already be obstructed **downstream** from the input ramp, e.g. by another input ramp. In this case ramp metering does not help the freeway, and only penalizes the supplying roadways.
- The ramp input may be getting its supply from another freeway, which may end up being jammed because of spillback, as discussed above.

We conclude that, more often than not, the freeway with its ramps and possibly other freeways constitute yet another store-and-forward system, and they should be treated as such. The optimization procedure should consider not just a portion of the freeway in the vicinity of the entrance ramp, but the entire freeway and its traffic-feeding partners, be they surface streets or other freeways.

3.8 SYSTEMS AFFECTED BY GEOMETRIC DETAILS

As mentioned already, moving and phantom bottlenecks can cause congestion at any point on a surface street or a freeway, thus becoming, in effect, temporary TPLs, albeit moving ones. Particularly notable example are the tunnels under the Hudson River, connecting New York City with New Jersey, which have received a great deal of attention by traffic theorists, from the early days of Traffic Theory. Examples are the contributions cited in references 57-63.

We shall discuss here the Lincoln tunnel, which demonstrates well the specific nature of the problem, and which was the subject of greater experimentation with control measures aiming at improving the efficiency of its operation. The key feature of this tunnel causing congestion is also found in other roadways, making the tunnel experience relevant to congestion control in general. Also common to all traffic systems is the challenge of measuring traffic densities, for which key contributions originated in the course of experiments aiming at improving the tunnel performance.

The Lincoln tunnel has a special geometry shown in Fig. 31. There are three distinct sections of the tunnel, a downgrade section, a level section, and an upgrade section. Generally, it is the upgrade portion that causes problems, as soon as demand increases into levels of congestion.

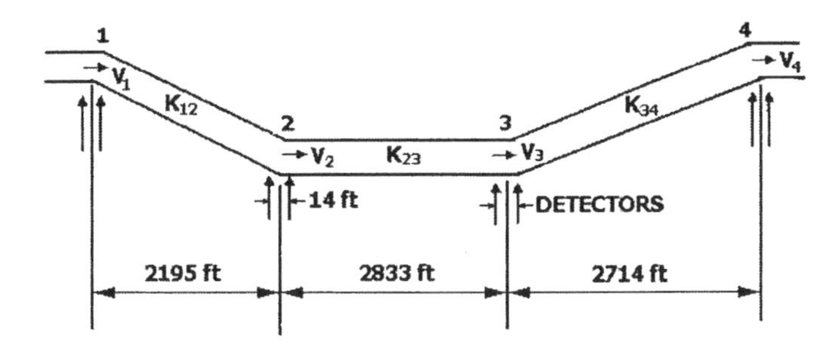

Figure 31. Geometric configuration of the South Tube of Lincoln Tunnel

As long as traffic densities inside the tunnel are low, traffic moves relatively freely, at flow rates approaching 1400 cars per lane per hour. However, congestion sets in as soon as demand increases and many cars pile into the tunnel. Congestion generally starts at the foot of the upgrade portion, point 3, as reported by Edie and Foote (1958). There, some cars, particularly trucks, can get into trouble. If they have to slow down just at point 3, they have difficulty accelerating, opening up a large gap in front and squandering throughput. They, in effect, become "moving bottlenecks" of a very special type, restricting traffic movements only in their lane, since lane changing is forbidden in the tunnel. The phenomenon is not unlike the "acceleration hysteresis" phenomenon described in Chapter 1, caused by the asymmetry between acceleration and deceleration of cars. The result of the moving bottleneck in the case of the tunnel is that its throughput is suddenly reduced to 1200 cars per lane per hour, or even less than that.

The cause of congestion was recognized for many years, and several attempts were made to mitigate it, including the work of Greenberg and Daou (1960), Herman and Potts (1961), Edie et al (1963), Crowley and

Greenberg (1965), and Foote (1968). In all these attempts, an effort was made to prevent congestion by limiting the input rate into the tunnel, often citing the paradoxical objective of "increasing throughput by decreasing input". The rule of thumb, which evolved during these attempts, was that limiting the input rate to 22 cars per minute, or 1320 cars per hour, could maintain relatively free flow and prevent congestion. The rate of 1320 cars per hour was below the maximum observed at the tunnel, but it often did work in preventing congestion for a few minutes. Unfortunately, this "open loop" control failed if, by chance a truck found itself at the foot of the upgrade at low speed and caused the beginning of congestion, after which the limited input became too high to handle. Eventually, "closed loop" control was implemented, first using some fixed-logic control hardware, (Foote and Crowley , 1967), and then an on-line digital computer, (Gazis and Foote, 1969). We discuss here this last experiment of computerized surveillance and control of the tunnel, which achieved the most reliable results, while also contributing to such areas as the measurement of traffic densities in sections of a roadway.

3.8.1 Density Estimation

Before any traffic control is implemented, it is imperative to know where the cars are. Thus, good surveillance is essential for the success of any traffic control measures, particularly for congested systems. It certainly is a prerequisite for the proper control of a store-and-forward system. Surveillance at the Lincoln tunnel certainly qualified as excellent, even though it used obsolescent technology by today's standards.

As shown in Fig. 31, four observation points, also referred to as "traps", were installed along the tunnel, at the entrance, foot of the downgrade, foot of the upgrade, and exit. The observation was done by photocells detecting the passage of a car over each one of the detectors of the trap, with a time resolution of about 10 msec. Typical profiles obtained by the photocell detectors are shown in Fig. 32, where the "car with variable number of holes" was generally a truck, with gaps seen by one but not both detectors, and the "VW pulling a ladder" was a truck pulling an empty platform. These last two signals were smoothed out logically to produce pulses similar to those of an ordinary car. The four times T_i shown in the figure were sufficient to provide a determination of the length, speed and (constant) acceleration of a vehicle passing over the trap. Speed and length could be obtained with great accuracy, and acceleration with a lesser one, but the acceleration was not used in the subsequent computations.

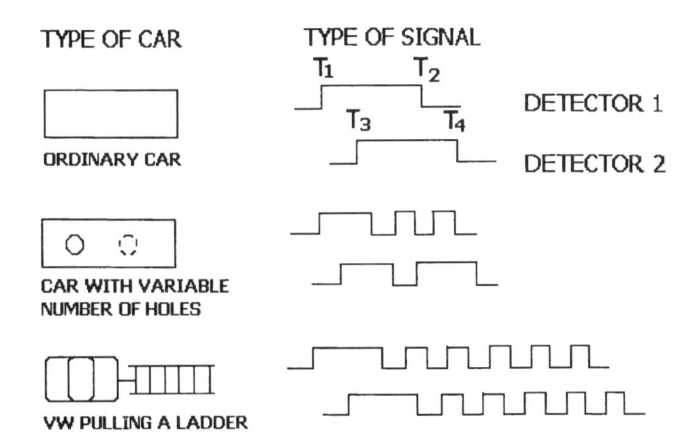

Figure 32. Types of signals from various types of vehicles

The speed of vehicles, together with the count of vehicles per unit of time, were used to determine the flow parameters at the observation point. The average speed is one of these parameters. Density, equal to the number of cars in a section divided by its length, could be obtained by initializing the count of cars somehow, and then using the formula

$$\left(K_{ij}\right)_{n+1} = \left(K_{ij}\right)_{n} + \left(N_{i}\right)_{n+1} - \left(N_{j}\right)_{n+1} \quad , \tag{94}$$

where $\left(K_{ij}\right)_{n}$ is the number of cars in section ij at the end of the n th "observation period", and $\left(N_{i}\right)_{n+1}$ and $\left(N_{j}\right)_{n+1}$ are the cars passing over the traps *ii* and *jj* during the (n+1)th observation period. Observation periods were equal segments of time, small enough to allow timely updating of the flow characteristics.

Equation 94 requires some initialization of the system with the count of vehicles in a section at time zero. After such an initialization, the correct count should be obtainable by this equation. Unfortunately, errors in the counts $\left(N_{i}\right)_{n+1}$ and/or $\left(N_{j}\right)_{n+1}$ would degrade the accuracy of this count. For this reason, correct updating of the count was necessary, periodically. Both, initialization and updating were done by identifying individual vehicles as they passed, first over the entrance trap of the section and then over the exit trap.

The lengths of vehicles were used to identify individual vehicles, or combinations of vehicles, such as a short vehicle following a long vehicle. When these patterns were repeated at the exit trap, the count of vehicles entering the tunnel section between the times of entrance and exit of the pattern used was the exact number of vehicles in the section at the time of exit of the pattern. The reader will recognize that in present days, with the availability of in-vehicle devices such as the RFID (Radio Frequency ID) used for electronic toll payment, the detection of individual cars is a very simple matter.

One may ask how densities can be estimated if identification of individual vehicles is not available. The Lincoln tunnel experiment provided an answer to this question. Having the accurate counts of vehicles at the Lincoln tunnel made possible the development and testing of other advanced, density estimation methods, such as the "Kalman Filtering" method, which is discussed in an appendix of this book.

3.8.2 The Control Algorithm

The detectors in the tunnel yielded the speeds and throughputs over the traps, and the counts of vehicles between the traps. On the basis of these variables, a control function could be developed, which would in general depend on the time series of these variables. Only one control action was available, namely, a restriction of traffic input into the tunnel. Initially, a binary decision was made between imposing or not imposing a restriction. Later, a four-level control possibility was implemented, refining the response to the measurements of speeds and densities over the entire tunnel

3.8.2.1 Binary Control, Single-Density Algorithm

The binary control decision was based only on the measurements of speed, V_3 , and traffic count, K_{23} , in the middle section of the tunnel, 23, (Fig. 31). The rationale was that it was in this section that the "acceleration hysteresis" initiated, by slowing trucks to almost a stop at the foot of the upgrade. "Observations" of the traffic were done at 5 sec. Intervals, and a control value C_n was determined for the nth observation interval, with a value equal to 0 or 1, the zero corresponding to absence, and the one to imposition of control. The control consisted of restricting the input of traffic into the tunnel in order to establish free flow, if possible. The control algorithm was the following:

- If $V_3 > 50$ fps, set $C_n = 0$;
- Otherwise, C_n is given by

$$C_n = H\left[K_{23} - A - BC_{n-1}\right] , \tag{95}$$

where H is the step function defined by

$$H(x) = \begin{cases} 0 & if \quad x < 0, \\ 1 & if \quad x \geq 0, \end{cases} \tag{96}$$

and A,B are constants allowing a variable threshold density for the imposition and removal of traffic input control. The values of A and B most frequently used were 25 and 1.

The single density, binary control algorithm worked some of the time, but often failed to anticipate congestion in time. In addition, the binary nature led to over-reaction or under-reaction at times. Or this reason, a multi-level control algorithm, based on the measurements at all three sections of the tunnel, was developed and implemented.

3.8.2.2 The Three-Density Algorithm

It is to be expected, and confirmed by observation, that congestion in any part of the tunnel can degrade its performance sooner or later. Congestion was generally defined as the case when speeds were lower than 20 mph, and densities higher than 55 veh/mile. A more accurate description of the effect of congestion in tunnel sections was developed through observation, and led to the following algorithm.

Let the state of the tunnel be described by three state variables, the counts K_{12}, K_{23}, and K_{34}. Then, the probability that congestion will set in within a short period, say 30 sec., increases with increasing values of any or all the K_{ij}. The control algorithm was based on the tacit assumption that the instantaneous values of K_{ij} were the primary determinants of impending congestion. In addition, four possible levels of control were implemented, corresponding to integer values for the control variable C ranging from 0 to 4. The value 0 corresponded to no control, and the values 1 to 4 corresponded to increasing restraints on entering traffic, with a red light corresponding to the value 4.

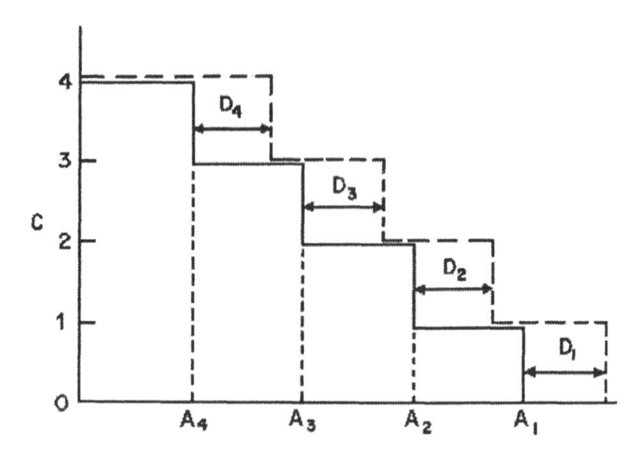

Figure 33. Threshold levels of the control function C, Eq. 98

A table was compiled in state space in state space, K_{ij}, which preceded congestion by about 30 sec., while the tunnel was under no control, or under single-density control. A smooth surface was fitted through these "critical density" points, leading to the following algorithm:

- If V_2 and V_3 re measured in fps and $V_2 V_3 > 2500$, set $C = 0$,
- Otherwise, compute

$$F = F(K_{ij}) = (P_{12} = K_{12})(P_{23} = K_{23})(P_{34} - K_{34}), \qquad (97)$$

and set

$$C = \sum_{k=1}^{4} \{1 - H[A_k + \mu D_k - F]\} \quad , \qquad (98)$$

where A_k, D_k, and P_{ij} are constants, H is the step function given in Eq. 96, and μ is a history-dependent variable, taking the values 0 and 1, which was used to generate the double-ladder function shown in Fig. 33. The meaning of the double-ladder control is that used in many "governor-type" control devices, namely, imposing or increasing control at some critical value of the state variables, but removing or reducing control at lower values of these variables, to avoid "chattering" of the control process. Thus, the value of μ was taken as 0 in computing ascending values of C, and 1 in computing descending values.

Application of the preceding control algorithm yielded a consistent improvement of the operation of the tunnel, with the average throughput increased by about 12%, compared to the uncontrolled state. This result provided proof that prevention of the "acceleration hysteresis" on roadways can be very beneficial. It may be worth applying on highways prone to such a hysteresis phenomenon, for example, a highway with a sudden steep uphill portion and no extra lane for slow-moving vehicles.

References

Allsop, R. E. (1968a), "Choice of offsets in Linking Traffic Signals", *Traffic Engr. & Control*, **10** , 73-75.

Allsop, R. E. (1968b) "Selection of Offsets to Minimize Delay to Trafic in a Network Controlled by Fixed-Time Signals", *Transport. Sci.*, **2**, 1-13.

Bavarez, E. and Newell, G. F. (1967), "Traffic Signal Synchronization on a One-Way Street", *Transport. Sci.*, **1**, 55-73.

Bellman, R. and Dreyfus, S. E. (1962), *Applied Dynamic Programming*, Princeton, NJ: Princeton Univercity Press.

Boillot, F., Blossville, J. M., Lesort, J. B., Motyka, V., Papageorgiou, M. and Sellam, S. (1992), "Optimal Signal Control of Traffic Networks", *6th Intern. Conf. On Road Traffic Monitoring and Control*, London, England, pp. 75-79.

Buckley, D. J., Hackett, L. G., Keuneman, D. J. K. and Beranek, L. A. (1966), "Optimum Timing for Coordinated Traffic Signals", *Proc. Aust. Road Res.Board,* **3**, Pt. 1, 334-353.

Crowley, K. W. and Greenberg, H. (1965), "Holland Tunnel Study Aids Efficient Increase of Tube's Use", *Traffic Eng.*, **35**, 20.

D' Ans, G. C. and Gazis, D. C. (1976), "Optimal Control of Oversaturated Store-and-Forward Transportation Networks", *Transportation Science*, **10**, 1-19.

Davison, E. J. and Ozguner , U. (1983), "Decentralized Control of Traffic Networks", *IEEE Trans. on Automatic Control*, **28**, 677- 688.

Diakaki, C., Papageorgiou, M. and McLean, T. (1999), "Application and Evaluation of the Integrated Traffic-Responsive Urban Corridor Control Strategy", *78th Annual Meeting of Transp. Res. Board*, Washington, DC, Paper No. 990310.

Dunne, M. C. and Potts, R. B. (1964), "Algorithm for Traffic Control", *Operations Res.*, **12**, 870-881.

Edie, L. C. and Foote, R. S. (1958), "Traffic Flow in Tunnels", *Highway Res. Board Proc.*, **37**, 334-344.

Edie, L. C., Foote, R. S., Herman, R. and Rothery, R. W. (1963), "Analysis of Single-Lane Traffic Flow", *Traffic Eng.*, **33**, 21-27.

Evans, H. K., Editor, (1950), *Traffic Engineering Handbook*, Institute of Traffic Engineers, New Haven, CT, p. 224.

Farges, J-L., Henry, J- J. and Tufal, J. (1983), "The PRODYN Real-Time Algorithm", *4th Intern. Symp. on Transp. Systems*, Baden Baden, Germany, pp. 307-312.

Foote, R. S. and Crowley, K. W. (1967), "Developing Density Controls for improved Traffic Operations", *Highway Res. Board Record*, **154**, No. 1430.

Foote, R. S. (1968), "Instruments for Road Transportation", *High Speed Ground Transp. Journal*.

Gartner, N. H. (1972a), "Constraining Relations among Offsets in Synchronized Signal Networks", Letter to the Editor, *Transport. Sci.*, **6**, 88-93.

Gartner, N. (1972b), "Optimal Synchronization of Traffic Signal Networks by Dynamic Programming", *Proc. 5th Intern. Symp. on the Theory of Traffic Flow and Transportation*, edited by G. F. Newell, New York: Elsevier, pp. 281-295.

Gartner, N. H. (1983), "OPAC: A Demand-Responsive Strategy for Traffic Signal Control", *Transp. Res. Record*, **906**, 75-84.

Gartner, N. H. (1991), "Road Traffic Control: Progression Methods", in *Concise Encyclopedia of Traffic and Transportation Systems*, edited by M. Papageorgiou, Pergamon Press, Oxford, UK, 391-396.

Gartner, N. H., Assmann, S. F., Lasaga, F. and Hom, D. L. (1991), "A Multiband Approach to Arterial Traffic Signal Synchronization", *Transp. Res.* **25B,** 55-74.

Gazis, D. C. (1964), "Optimum Control of a System of Oversaturated Intersections", *Operations Res.*, **12**, 815-831.

Gazis, D. C. (1965a), "Traffic Control, Time-Space Diagrams and Networks", *Traffic Control – Theory and Instrumentation*, edited by T. R. Horton, New York: Plenum, pp. 47-63.

Gazis, (1965b), "Spillback from an Exit Ramp of an Expressway", *Highway Research Record*, **89**, 39-46.

Gazis, D. C. and Potts, R. B. (1965), "The Over-Saturated Intersection", *Proc. 2nd Intern. Symp. on the Theory of Road Traffic Flow*, edited by J. Almond, Paris: Organization for Economic Cooperation and Development, pp. 221-337.

Gazis, D. C. and Bermant, O. (1966), "Dynamic Traffic Control Systems and the San Jose Experiment", *Proc. 8th Intern. Study Week in Traffic Engineering*, Barcelona, Spain, **V**.

Gazis, D. C. and Foote, R. S. (1969), "Surveillance and Control of Tunnel Traffic by an On-Line Digital Computer", *Transp. Science*, **3**, 255-275.

Grafton, R. B. and Newell, G. F. (1967), "Optimal Policies for the Control of an Undersaturated Intersection", *Proc. 3rd Intern. Symp. on Traffic Theory*, edited by L. C. Edie, R. Herman, and R. W. Rothery, New York: Elsevier, pp.239-257.

Greenberg, H. and Daou, A. (1960), "The Control of Traffic Flow to Increase the Flow", *Operations Research*, **8**, 524-532.

Herman, R. and Potts, R. B. (1961), "Single-Lane Traffic Theory and Experiment", *Theory of Traffic Flow*, edited by R. Herman, Amsterdam: Elsevier Publishing Company, pp. 120-146.

Hillier, J. A. and Lott, R. S. (1966), "A Method of Linking Traffic Signals to Minimize Delay", *Proc. 8th Intern. Study Week in Traffic Engineering*, Barcelona, Spain, **V**.

Holroyd, J. and Hillier, J. A. (1969), "Area Traffic Control in Glasgow – A summary of Results from Four Control Schemes", *Traffic Eng. & Control*, **11**, 220-223.

Hunt, P. B., Robertson, D. L. and Bretherton, R. D. (1982), "The SCOOT on-line traffic signal optimization technique", *Traffic Eng. & Control*, **23**, 190-192.

Kane, J. N. (1964), *Famous First Facts*, New York: H. W. Wilson, 3rd ed., p. 626.

Katz, S. (1962), "A Discrete Version of Pontryagin's Maximum Principle", *J. Electron. Control*, **13**, No. 2, 8.

Kessaki, A., Farges, J-L. and Henry, J-J. (1990), "Upper Level for Real Time Urban Traffic Control Systems", *11th IFAC World Congress*, Tallinn, Estonia, Vol. 10, pp.226-229.

Kim, K.-J., and Bell, M. G. H. (1992), "Development of an Integrated Traffic Control Strategy for Both Urban Signalized and Motorway Networks", *Proc. 1st Meeting of the EURO Working Group on Urban Traffic and Transportation*, Landshut, Germany.

Koshi, M. (1972), "On-Line Feedback Control of Offsets for Area Control of Traffic", *Proc. 5th Intern. Symp. on the Theory of Traffic Flow & Transportation*, edited by G. F. Newell, New York: Elsevier, pp.269-280.

Kwakernaak, H. (1966), "On line Optimization of Stochastic Control Systems", *Proc. 3rd Intern. Federation of Automatic Control*, London.

Lim, J. H., Hwang, S. H., Suh, I. H. and Bien, Z. (1981), "Hierarchical Optimal Control of Oversaturated Urban Networks, *Intern. Journal of Control*, **33**, 727-737.

Little, J. D. C. (1964), "The synchronisation of Traffic Signals by Mixed-Integer Linear Programming", *Operations Res.*, **14,** 568-594.

Luk, J. Y. K. (1984), "Two Traffic-Responsive Area Traffic Control Methods: SCAT and SCOOT", *Traffic Eng. & Control*, **25**, 14-22.

Marsh, B. W. (1927), "Traffic Control", *Ann. Amer. Soc. Political Social Sci.*, **133**, 90-113.

Matson, T. M., Smith, W. S. and Hurd, F. W. (1955), *Traffic Engineering,* McGraw-Hill Book Company, Inc., New York.

Michalopoulos, P. G. and Stephanopoulos, G. (1977a), "Oversaturated Signal Systems with Queue Length Constraints – 1: Single Intersection", *Transp. Research*, **11**, 413-421.

Michalopoulos, P. G. and Stephanopoulos, G. (1977b), "Oversaturated Signal Systems with Queue Length Constraints – 2: System of Intersections", *Transp. Research*, **11**, 423-428.

Miller, A. J. (1965), "A Computer Control System for Traffic Networks", *Proc. 2nd Symp. on Traffic Theory*, edited by J. Almond, Paris: Organization for Economic Cooperation and Development, pp.200-220.

Okutani, I. (1972), "Synchronization of Traffic Signals in a Network for Loss Minimizing Offsets", *Proc. 5th Intern. Symp. on the Theory of Traffic Flow and Transportation*, edited by G. F. Newell, New York: Elsevier, pp. 297-312.

Papageorgiou, M., Haj-Salem, H. and Blossevil, J.-M. (1991), "ALINEA: A Local Feedback Control Law for On-Ramp Metering", *Transp. Res. Record*, **1320**, 58-64.

Papageorgiou, M. (1999), "Traffic Control", in *Handbook of Transportation Science*, Boston: Kluwer Academic Publishers, pp. 233-267.

Park, E. S., Lim, J. H., Suh, I. H. and Bien, Z. (1984), "Hierarchical Optimal Control of Urban Traffic Networks", *Intern. Journal of Control*, **40**, 813-829.

Pontryagin, L. S., Boltyanskii, V. G., Gamkrelidze, R. V. and Mishchenko, E. F. (1962), *The Mathematical Theory of Optimal Processes*, translated by K. N. Trirogoff, New York: Wiley-Interscience.

Rathi, A. K. (1988), "A Control Scheme for High Density Sectors", *Transp.* Research, **22B**, 81-101.

Robertson, D. I. (1969), " 'TRANSYT' Method for Area Traffic Control", *Traffic Engr. & Control* , **11**, 276-281.

Ross, D. W., Sandys, R. C. and Schlaefli, J. L. (1970), *A Computer Control Scheme for Critical-Intersection Control in an Urban Network*, Menlo Park, CA, Stanford Research Institute.

Ross, D. W. (1972), "Traffic Control and Highway Networks", *Networks*, **2**, 97-123.

San Jose Traffic Control Project – Final Report, (1966), San Jose, CA: IBM Corp., Data Processing report.

Sen, S. and Head, L. (1997), "Controlled Optimization of Phases at an Intersection", *Transp. Science*, **31**, 5-17.

"SIGOP", Traffic Research Corporation, New York, 1966. Distributed by Clearinghouse for Federal Scientific and Technical Information, **PB** 17 37 38.

Singh, M. G. and Tamura, H. (1974), "Modelling and Hierarchical Optimisation of Oversaturated Urban Traffic Networks", *Intern. Journal of Control*, **20**, 269-280.

Van Zijverden, J. D. and Kwakernaak, H. (1969), "A New Approach to Traffic-Actuated Computer Control of Intersections", *Proc. 4th Intern. Symp. on Traffic Theory*, edited by W. Leutzbach and P. Baron, Bonn: Bundesminister fur Verkehr, pp.113-117.

Webster, F. V. (1958), "Traffic Signal Settings", Road Research Technical Paper No. 39, Road Research Laboratory, England.

Additional references

Allsop, R. B. (1971), "SIGSET: A Computer Program for Calculating Traffic Capacity of Signal-Controlled Road Junctions", *Traffic Eng. And Control*, **12**, 58-60.

Allsop, R. B. (1976), "SIGCAP: A Computer Program for Assessing the Traffic Capacity of Signal-Controlled Road Junctions", *Traffic Eng. And Control*, **17**, 338-341.

Boillot, F., Blosseville, J. M., Lesort, J. B., Motyka, V., Papageorgiou, M. and Sellam, S. (1992), "Optimal Signal Control of Traffic Networks", *6th IEE Intern. Conf. On Road Traffic Monitoring and Control*, London, England, pp.75-79.

Elloumi, N., Haj-Salem, H. and Papageorgiou, M. (1996), "Integrated Control of Traffic Corridors – Application of an LP Methodology", *4th Meeting of the EURO Working Group on Transportation Systems*, New Castle, UK.

Farges, J-L, Henry, J-J. and J. Tufal, J. (1983), "The PRODYN Real-Time Traffic Algorithm", *4th IFAC Symp. on Transp. Systems*, Baden Baden, W. Germany, pp. 307-312.

Gervais, E. F. (1964), "Optimization of Freeway Traffic by Ramp Control", *Highway Res. Rec.*, **59**, 104.

Haj-Salem, H. and Papageorgiou, M. (1995), "Ramp Metering Impact on Urban Corridor Traffic – Field Results", *Transp. Res.*, **29A**, 303-319.

Hillier, J. A. (1966), "Appendix to Glasgow's Experiment in Area Traffic Control', *Traffic Engr. & Control*, **7**.

May, A. D. (1964), "Experimentation with Manual and Automatic Ramp Control", *Highway Res. Rec.*, **59**, 9-38.

Messner, A. and Papageorgiou, M. (1995), "Automatic Control Methods Applied to Freeway Network Traffic", *Automatica*, **30**, 691-702.

Miller, A. J. (1963), "A Computer Control System for Traffic Networks", *Proc. 2nd Intern. Symp. on Traffic Theory*, London, UK, pp.200-220.

Moreno-Banos, J. C., Papageorgiou, M. and Schaeffner, C. (1993), "Integrated Optimal Flow Control in Traffic Networks", *Europ. J. of Oper. Res.*, **71**, 317-323.

Papageorgiou, M. (1983), *Applicaton of Automatic Control Concepts to Traffic Flow Modeling and Control*, New York: Springer Verlag.

Papageorgiou, M. (1998), "Automatic Control Metods in Traffic and Transportation", in *Operations Research and Decision Aid Methodologies in Traffic and Transportation Management*, P. Toint, M. Labbe, K. Tanczos, and G. Laporte Editors, New York: Springer Verlag, pp. 46-83.

Wattleworth, J. A. and Berry, D. S. (1965), "Peak-Period Control of a Freeway System – Some Theoretical Investigations", *Highway Res. Rec.*, **89**, 1-25.

Chapter 4

Traffic Generation, Distribution, and Assignment

One of the key prerequisites for proper management of transportation systems, including both transportation planning and traffic control, is knowing where traffic appears, and why it appears. This leads to the need for understanding how traffic is "generated", how it is "distributed" from a set of origins to a set of destinations, and how it is "assigned" along given routes.

Traffic generation is the result of the existence of people in some parts of a network, and the existence of attractions, such as jobs, in another part. Knowing the numbers of people and attractions allows one to determine how traffic is generated, and how it is distributed into an "origin-destination" table for the network. The generation and distribution of traffic may be estimated on the basis of historical data, or it may be based on probabilistic estimates of certain attractions, such as special events or sales. The traffic assignment issue concerns the way that these origin-destination needs are met by selecting the routes to be taken by individual drivers.

The investigation of all three topics has borrowed from Graph Theory of Flows in Networks, (see Ford and Fulkerson, 1962), which has been developed over the years for the investigation of networks of all types, and in particular communication networks. In what follows, we shall discuss the network representation of a transportation system, and the methodology for investigating the traffic generation, distribution, and assignment for such systems.

4.1 NETWORK REPRESENTATION OF A TRANSPORTATION SYSTEM

An obvious example of a transportation network is an urban street grid such as that shown in Fig.1. In Graph Theory nomenclature, the intersections of such a network are the *nodes*, and the street segments between nodes are the *links* of the network.

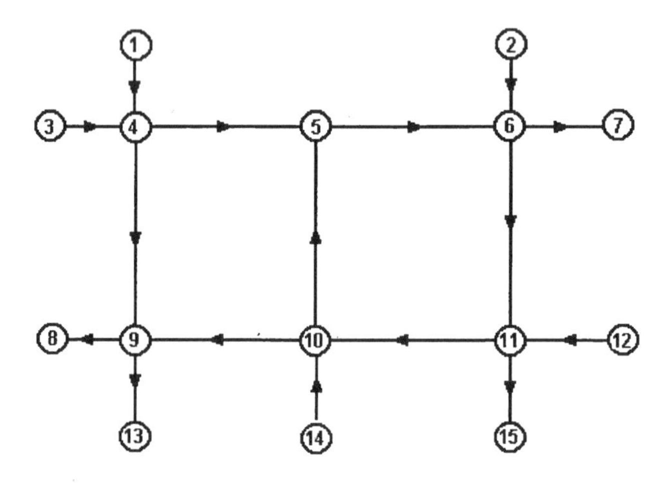

Figure 1. A network of intersections and street segments

If we wish to account for turning movements as well, the network around the intersections will include additional links. For example, the network around intersection 4 of Fig. 1 will be altered to one shown in Fig. 2, with node 4 replaced by the nodes 16 to 19. The throughput of the links in networks such as those in Figs. 1 and 2 are selected in such a way as to include the effects of traffic lights at the various intersection nodes.

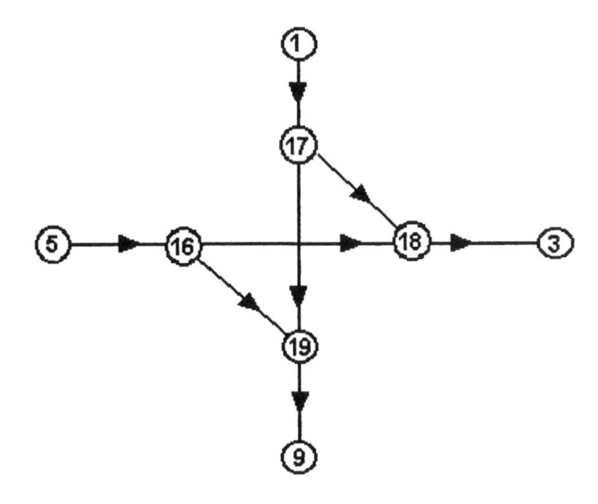

Figure 2. A network representation of turning movements

Networks of intersections such as those of Figs. 1 and 2 may be occasionally used for the study of regions of critical intersections. However, for a more global transportation planning, it is customary to decrease the granularity of the network representation.

An urban region is divided into a few *sectors*, then into smaller *districts*, and finally into *zones*, such as the 15 zones shown in Fig. 3. Traffic is viewed as moving from one zone to another in the network. It is customary to view traffic as emanating from a *centroid* of the node, or reaching the centroid if originating from another centroid. All discussions of traffic generation, distribution and assignment refer to movements of traffic between the centroids plus the entering and exiting odes of the network. The centroids are connected either to the nodes of the network, or to some mid-points along the links. The network itself may either be a geometric representation of actual roadway segments and their intersections, or it may be the result of an aggregation of roadway segments within zones into a network representing traffic movements from zone to zone. The *Spider Web* network, in which nodes are the centroids, and links connect adjacent

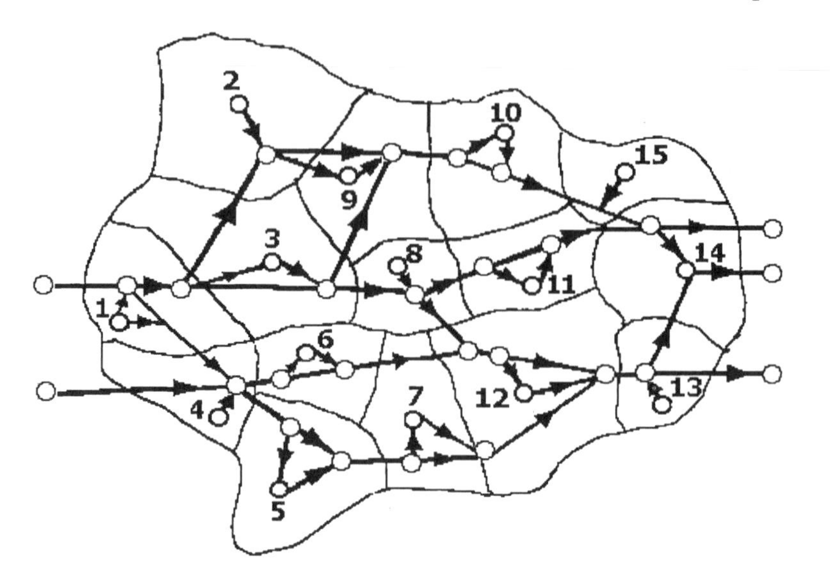

Figure 3. A network representation of urban traffic movements

centroids, is of the latter type, in which the links do not correspond to actual road segments, but represent desired movements of traffic between zones. If the network representation incorporates actual roadway sections, then the properties of these sections can be taken into account explicitly in estimating such things as travel times between nodes, or between centroids. If a spider web network is used, then travel times along its links are estimates of actual travel times between zones using all available roadway choices. In any case, a complete description of a link would provide travel time and/or cost as functions of the link occupancy, rate of flow, time, and other pertinent factors. A node represents the terminus of one or more links, and may be used either to join the termini of a number of links, or to represent a *source* (origin) or *sink* (destination) of traffic. In Fig. 3, the numbered nodes are the centroids, and the unnumbered ones are nodes of the roadway network.

Let the nodes of the network be labeled $1, 2, 3, \ldots\ldots, N$. It is assumed that all links are *directed*, meaning that they can only be traversed in one direction. A directed link from node i to node j is represented by (i,j). If a network segment between nodes ii and jj carries traffic in both directions, it is represented by two links, (i,j) and (j,i), as shown in Fig. 4, preserving the directed graph quality of the network.

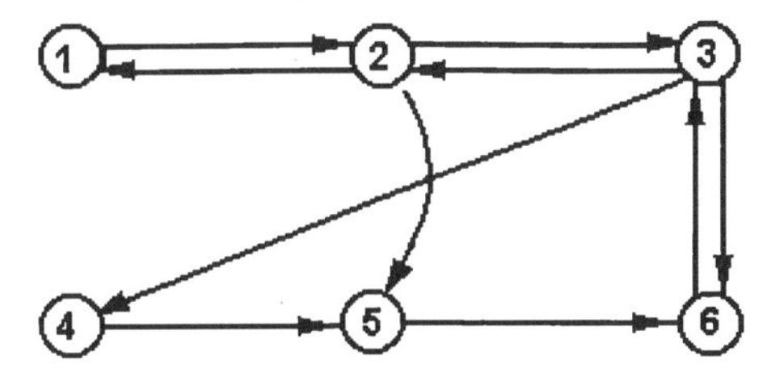

Figure 4. Example of a network graph.

The preceding labeling strategy allows one to describe the graph fully by listing all the links of the graph. For example, the network of Fig. 4 is described as comprising the links (1,2), (2,1), (2,3), (3,2), (2,5), (3,4), (3,6), (6,3), (4,5), ad (5,6). A *chain* from node i to node j is a specified route comprising links directed the same way, thus permitting travel from i to j. A *cycle* is a chain that comes full circle and terminates at its origin node. A *path* between nodes ii and jj is a sequence of links between these two nodes, not all of which are necessarily directed the same way. In Fig. 4, (1,2)(2,3)(3,6) , (12)(2,5)(5,6) , and (1,2)(2,3)(3,4)(4,5)(5,6) are chains from 1 to 6, and (2,5)(5,6)(6,3)(3,2) is a cycle from 2 to itself. A path from node 4 to node 1, which is not a chain, is (4,5)(2,5)(2,1), which is also a path from 1 to 4.

A node j is *accessible* to node i if there exists a chain from i to j. A *connected network* is one in which there are paths between all pairs of nodes. The network of Fig. 4 is such a connected one. A *tree* is all or part of a network which is connected but contains no cycles. For example, the network of Fig. 4 would become a tree if we removed the links (3,6) and (6,3). The above definitions will suffice for the purpose of the discussion in this book. A more detailed description of graphs can be found in Ford and Fulkerson (1962) , for a mathematician's view, and in Potts and Oliver (1972), for the vehicular traffic application.

4.2 TRIP GENERATION AND DISTRIBUTION

In plȧnning, or trying to improve the management of a traffic network, one needs a good estimate of the traffic load carried by the current state of the network. This means that one needs to know how traffic is *generated* at any node, how it is *distributed* by apportioning of trips originating at that node among all possible destinations, and how one route from origin to destination is *assigned* to any trip. The computed values and choices for these three processes certainly change with the time of the day. They may also change from one day to the next, in response to changing conditions, (e.g. construction), or special attractions. They certainly change over long periods of time, due to changes in the distribution of population, jobs, and shopping and entertainment opportunities, to mention some of the primary factors of influence.

Clearly, these three processes are interrelated. For example, if a trip is generated because of a special attraction at a particular node of the network, it will be natural to conclude that the newly generated trips will involve their distribution into the attraction node. Some of the more recent work in this area does consider the generation and distribution of traffic together. Let us outline first some of the elements of these processes.

4.2.1 Trip Generation

Trip generation depends on the number of people living in a particular zone, and their travel needs for various purposes, such as jobs, shopping, or entertainment. For a first approximation, historical values of trip generation may be used as a basis, to which may be added expected effects of current attractions. For a more systematic approach to the task, particularly for the purpose of obtaining trip generation estimates over a relatively long horizon, it is customary to view the process in two stages:

1. The prediction of trip generation prediction variables, such as population, car ownership, shopping opportunities, and job availability at various nodes, and
2. The prediction of the number of trips originating from a zone, generated by a particular trip purpose, as a function of the predictor variables.

It is generally assumed that all travel consists of a trip from a particular origin to a destination, and a return trip on the same day. This is obviously an approximation, since it ignores multi-destination travel, and trips not generating a return trip on the same day, but for most cases of urban traffic the approximation is a reasonably good one. Basic data needed for estimating some of the key predictor variables, such as population, economy, and car ownership, may be obtained through direct traveler surveys, and/or publicly available census data. A full discussion of the processes involved in trip generation and distribution can be found in Helly (1974).

4.2.1.1 Prediction of the Predictor Variables

POPULATION. The number of people living in a particular zone can be changed by births, deaths, and *net migration* into the zone, i.e. the difference between the number of people moving into a zone and the number of people moving out of the zone. Birth and death rates may be assumed according to historical data. Net migration, ideally, should be correlated to various driving factors, such as quality of neighborhoods, proximity to jobs and attractions, etc. For a first approximation, it is customary to simply extrapolate the graphical or mathematical representation of past changes. Such extrapolations may take into account the nature of a neighborhood in terms of history and development. An old, established neighborhood is likely to have a relatively constant rate of population change, resulting in an exponential increase or decrease with time, possibly tending asymptotically to some limit value. A newly developing neighborhood may have various stages of development. For example, it may start with an early period of slow population growth, followed by a middle stage of rapid population increase, and ending with a third stage of exponential/asymptotic growth similar to that of an established neighborhood.

In estimating population changes, it is important to take into consideration expected unusual changes in conditions, such as a large change in employment opportunities due to externally imposed conditions, such as government initiatives. Also, estimates of population growth may involve a region comprising more than one zone, in which case the estimate for a single zone is obtained as a fraction of the regional estimate, taking into account historical data regarding the value of such a fraction.

Sometimes it is appropriate to estimate population as a function of future employment, particularly in the case of an area where employment is likely to change drastically as a result of an infusion of investment. Frequently, government funds may be the driving force for such an unusual growth of employment. It may be noted that the *labor force participation ratio,* (the ratio of employment to the total population), may be abnormally high in an area with very rapid employment growth. Furthermore, this ratio appears to be rising constantly, though slowly, in the United States, and it is hovering above 0.4 at the present.

ECONOMY. The economy of a region clearly affects travel in general, and travel between nodea of a network, in particular. Economic forecasts can, and often are made on the basis of an extrapolation from past historical data. It is generally assumed that the local economy depends on some principal economic sectors, which are those whose sales are largely exported to points outside the study area, and whose scale of operation may be extrapolated on a regional or national basis. It is postulated that changes in these basic sectors lead to changes in the rest of the local economy, affecting employment and other factors pertinent to the traffic generation issue.

A more scientific approach to the prediction of long-term economic development is known as the *Leontief input-output model*, which was used in a new York area study (see Berman et al, 1961), and in the Northeast Corridor Transportation Study, (see Putman, 1966). This model is based on the fact that the output of a sector must be equal to the sum of the inputs from this sector to all other sectors. It is specialized, for the purpose of deriving traffic generation data, by taking into account explicitly purchasing within an urban area of a portion of the products produced in that area. A reasonable extrapolation of industrial activity into the future yields the levels of activity of the various industrial sectors, which in turn drive the traffic generation process.

LAND USE. Clearly, the nature of land use influences the source and nature of travel, and consequently the traffic generation and distribution processes. Several approaches to predicting land use have been contributed, and partially tested. They all require extant land use inventories, and independent forecasts of the overall demand for land, the latter expressed in terms of building construction rates for the area. The models are either simple extrapolations of past trends, or simulations of assumed human behavior, both constrained by the available resources. The simulations may be further subdivided into deterministic models, in which people are guided only by accepted economic principles, or probabilistic models, in which human behavior is only partially based on such principles.

An example of a simple extrapolation involves developing a multiple linear regression equation of the form

$$Y = c_0 + \sum_{i=1}^{n} c_i X_i \ , \tag{1}$$

where Y is the predicted change in the zone's inventory of dwelling units, retail space, factory activities, and other land use measures, X_1 to X_n are independent predictor variables, such as vacant land area, land value, distance to employment sites or the centroid of the urban area, as well as other variables describing zoning regulations influencing density and types of construction, and availability of transit. The problem is to determine the constants c_0 to c_n on the basis of past trends. Having land use surveys for at least two, and preferably more, points in time allows such a determination.

One of the early applications of this approach was for the Greensboro, North Carolina area, by Swerdoff and Stowers (1966).

Another approach, which takes into account two different drivers of land use, is known as the *density-saturation gradient method.* It was used in the Chicago Area Transportation Study, (see Hansen, 1960), and it consists of adjusting for consistency two extrapolated curves. One is for dwelling unit density as a function of travel time to a central point, and the other is for saturation levels of land use, also as a function of travel time to the central point.

Two probabilistic simulation approaches, also discussed by Helly (1974), are the following:

The first, known as the *accessibility* or *gravity* model. Is based on the postulate that the likelihood of building on a site in a given zone depends mainly on the zone's distance from places of employment or commerce, which gives rise to trips originating from the study zone. The other, known as the *intervening opportunity* model, is based on the assumption that the likelihood of building on a given site is a monotonically decreasing function of the number of equivalent available sites which are nearer to the centroid of the attractions listed above, namely, employment and commerce. Since the predictors for both models change in time, time-dependent models have been developed for both.

The accessibility model, first used in a time-independent form by Hansen[7], has the form

$$g_i = \left(\sum_i g_i\right) \frac{A_i^a V_i}{\sum_i A_i^a V_i} , \tag{2}$$

where g_i is the forecast expected growth rate for zone i, A_i an accessibility index for zone i with an empirically determined constant exponent a, and V_i the vacant land area in zone i. The accessibility index has the general form

$$A_i = \sum_j E_j F_{ij} , \tag{3}$$

where E_j is a measure of activity in zone *j* such as total employment, and F_{ij} is a measure of "friction" between nodes *i* and *j*, which may be made proportional to the actual number of trips between the two nodes, as surveyed or predicted exogenously. If no data exist allowing the estimation of this friction, F_{ij} may be chosen as $F_{ij} = 1/T_{ij}^b$, where T_{ij} is the travel time between zones *i* and *j* , and *b* a constant exponent.

Intervening opportunity models have been proposed by Stouffer (1940), Schneider (1960), Sherratt (1960), and Helly (1969), the last one being a time-dependent one. The general approach to the intervening opportunity models is the following.

Let us assume that there is just one centroid, and a builder's site selection proceeds from the site nearest to the one farthest from the centroid. Let $M(r)$ be the number of sites within a radius r from the centroid, where because of the large number of sites in a city the integer-valued $M(r)$ is treated as a continuous variable. Let dM be the number of sites between M and $M+dM$, and $p\,dM$ the probability that a given house's site is selected in the range dM, given that it was not selected nearer to the centroid than M. The parameter p is assumed constant. Then, the probability of not selecting a site in the range $(0, M+dM)$ is given by

$$Pr\,(M + dM) = Pr\,(M)\,(1 - p\,dM)\,, \tag{4}$$

which has the time-independent solution

$$Pr\,(M) = e^{-pM}\quad . \tag{5}$$

If a total of N_T houses are built, the number built on the M most central sites is $N(M) = [\,1 - Pr\,(M)]$, and the density of houses, $n\,(M)$ is given by

$$n(M) = \frac{dN(M)}{dM} = pN_T\,e^{-pM}\quad . \tag{6}$$

The preceding time-independent formulation does not ensure, as it should, that $N(M) \leq M$. This deficiency can be overcome by a time-dependent reformulation in which a site count variable *m* is introduced, with all available sites at time zero described, in order of decreasing accessibility from the centroid, by the integers 1 to *m*. Again, because of the large number of sites in an urban area, the parameter *m* is treated as a continuous variable. Now, we can define $M(m, t)$ as the number of sites still vacant at time *t*, which are no less desirable than *m*. We note here that $M(m, 0) = m$, and that $M(m, t)$ has the same meaning as the time-independent *M*. Thus, Eq. 5 still gives the probability $Pr(M)$, except that *M* is now time-dependent.

Let $N(t)$ be the total number of sites occupied at time *t*, and therefore $M(m,t)$ be equivalent to $M(m,N)$. During a time increment *dt*, *dN* houses will be built. Of these, a fraction $[1 - Pr(M)]$ can be expected to be built on the *M* most accessible and still vacant sites. The number built during *dt* on these *M* sites equals the number that must be removed from the vacant site stock, therefore we have the relationship

$$\frac{dM(m,t)}{dt} = \frac{dM(m,N)}{dt} = -\frac{dN}{dt}[1 - Pr(M)] \,, \tag{7}$$

and using Eq. 5, we obtain

$$\frac{dM}{dN} = -\left(1 - e^{-pM}\right) \,, \tag{8}$$

which has the solution

$$M(m,N) = \frac{1}{p}\log\left[\left(e^{pm} - 1\right)e^{-pN} + 1\right] \,. \tag{9}$$

Now, let $F(m,N)$ represent the expected fraction of sites occupied in the range $(m, m+dm)$, which is found to be

$$F(m,N) = dm\frac{d[m - M(m,N)]}{} = \frac{e^{pN} - 1}{e^{pN} + e^{pm} - 1} \quad . \tag{10}$$

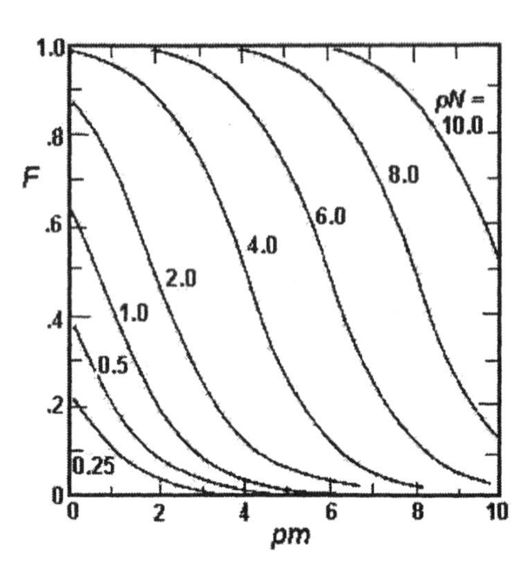

Figure 5. Housing site occupancy

Figure 5, from Helly (1974), shows the form of Eq. 10 for various values of pN and pm. In this figure, F is the probability that a site is occupied, m is the number of sites in decreasing order of accessibility, N is the total number of occupied sites, and p is the probability that a "considered" site is accepted. Since the abscissa is in units of pm, large values of it correspond to relatively inaccessible sites with large m. The curves in Fig. 5 show the progression of a city from its beginning, with few houses and pN small, to its later stages with increasing pN, when more and more desirable sites have been built upon, and new construction takes place in the outer fringes. Needless to say that this argument applies only to new construction and not rebuilding of earlier structures.

The above discussion has focused on residential land use, but similar models can be used for commercial or industrial developments. In such cases, we need data on the need for such developments, and knowledge of the often stringent constraints concerning such developments, either imposed by authorities or dictated by considerations of entrepreneurial success.

4.2.1.2 Trip Generation from individual zones

Having the values of the predictor variables, one can proceed to the prediction of trips originating from a zone. Trips are made by drivers and passengers of private cars, by people riding buses and taxis, and by commercial vehicles such as trucks. In many cities, private cars predominate, although this may not be the case in very large metropolitan areas like New York City. In any case, the trip generation picture is completed by translating the predicted data on population, economy, and land use, into individual trips.

It has been found that often, population density and household income are the key factors influencing trip generation from households, including commuting trips to work and other trips during the day. This is because automobile ownership is highly correlated with these two variables. Another factor that must be taken into account is the availability of *modal choice*, for example, using available public transportation for commuting to work.

For modeling purposes, the number of trips per household may be modeled by a correlation function of the predictor variables, fitted to observed facts in the general area of a zone. It is reasonable to assume that activity patterns do not change drastically overnight, and therefore an extrapolation of current activity patterns into the future is a reasonable approximation. However, as the planning horizon increases, account must be taken of changes in behavioral patterns and other factors which cause deviations from a simple extrapolation.

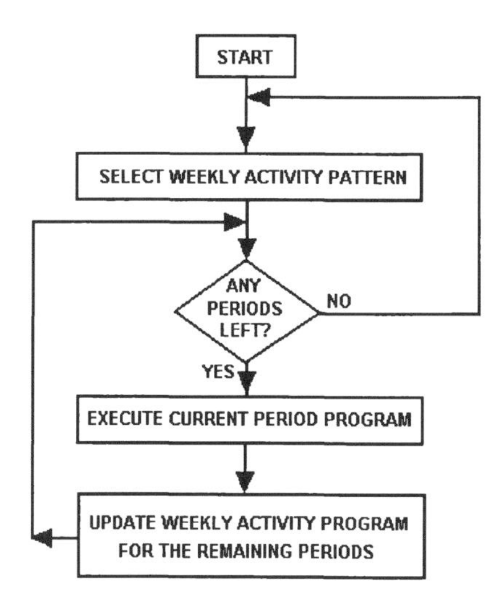

Figure 6. Determining weekly activities

Important contributions to the problem of trip generation have been made by modeling activity patterns in an urban area, which generate trips. The relevance of activity patterns was established by studies such as those of Stopher et al (1979), Pas (1982), Damm (1983), and Jones (1983). Hirsh et al (1986) contributed a dynamic, activity-based model of weekly activity model. In their model, it is assumed that the week is divided into several time periods. An individual selects an activity pattern at the beginning of the week, and updates the pattern at the beginning of each subsequent period on the basis of experiences during the preceding periods, and additional information acquired during those periods. Figure 6 shows the decision process for generating and updating a weekly activity program, following Hirsh et al. Having the activity pattern leads to the determination of the necessary trips for these activities.

4.2.2 Trip Distribution

Having the trip generation completed, the next step is to apportion the trips originating from any zone to all possible destinations. Various models used for this purpose are based on the levels of activity at the various destinations, and the difficulty in getting to these destinations. It has been found that the accuracy of trip distribution is improved if it is subdivided into categories of activities, such as work, shopping, business or school. Any model proposed must be validated and calibrated through the use of current trip data, whose acquisition is one of the key challenges in transportation planning.

In the models that are described below, it is assumed that an urban area is divided into N zones, which comprise the totality of the area of interest. Trips from outside the area are accounted for by being assigned to one of the peripheral nodes of the area network. The following notation is used:

t_{ij} = number of trips from zone i to zone j during a specified period, e.g. a weekday rush hour, as predicted by a trip distribution model for an observable period.

t_{Eij} = an estimate of t_{ij} obtained exogenously to the trip distribution model, by survey or observation.

$t_i = \sum_{j=1}^{N} t_{ij}$ = number of trips originating in zone i during the specified time period, as predicted by a model for an observable base period.

$t_{Ei} = \sum_{j=1}^{N} t_{Eij}$ = an estimate of t_i obtained exogenously to the trip distribution model.

T_{ij} = number of trips from zone i to zone j during the specified time interval, as predicted by the trip distribution model.

$$T_i = \sum_{j=1}^{N} T_{ij}$$

T_{Ei} = an estimate of T_I obtained exogenously to the trip distribution model, by application of a trip generation analysis.

If the period of interest is one day, then, on the basis of reasonable expectations about commuting patterns, it is assumed that $t_{ij} = t_{ji}$, $t_{Eij} = t_{Eji}$, and $T_{ij} = T_{ji}$. A matrix form representation of any of these quantities is a convenient form of their display.

Some of the models proposed for the treatment of trip distribution are known as *Growth Factor Models*, which try to create a set of origin-destination choices compatible with the growth of trips originating in various zones, and the growth of attractions in other zones. Other model use the *gravity* and *intervening opportunity* concepts discussed in Section 4.2.1 in conjunction with land use. Yet another approach to trip generation is provided by a variant of the gravity model known as the *minimum entropy* model.

4.2.2.1 Growth Factor Models

Let us define a *Growth Factor*, G_i, according to the expression

$$G_i = \frac{T_{Ei}}{t_{Ei}} = \frac{\textit{predicted future trips originating in zone i}}{\textit{observed present trips originating in zone i}} .$$

If we assume that trip distribution is done entirely on the basis of such growth factors for all the N zones, $G_i \left(i = 1,2,\cdots,N\right)$, then a simple approach to trip distribution is based on averaging the growth factors of origins and destinations, yielding

$$T_{ij} = t_{Eij}\left(\frac{G_i + G_j}{2}\right) \tag{11}$$

Admittedly, such a simple extrapolation ignores changes in travel parameters such as travel times, which are induced by growth, and can be used for relatively small time horizons. In any case, one must iterate in order to produce internally consistent results. The value of $T_i = \sum_{j=1}^{N} T_{ij}$ obtained on the basis of Eq. 11 is generally different from T_{Ei}, which is a basic estimate of traffic volume for zone i. In order to adjust the T_{ij} so that T_i comes close to T_{Ei}, we may view the value obtained from Eq. 11 as the first step of an iterative procedure, in which

$$\begin{aligned} T_{ij}^{(n+1)} &= T_{ij}^{(n)}\left(\frac{G_i^{(n+1)} + G_j^{(n+1)}}{2}\right) \\ G_i^{(n+1)} &= \frac{T_{Ei}}{\sum_{j=1}^{N} T_{ij}^{(n)}}, \quad n \geq 1 . \end{aligned} \tag{12}$$

The iteration is repeated until some step m for which

$$\left(T_{Ei} - T_i^{(m)}\right) = \left(T_{Ei} - \sum_{j=1}^{N} t_{ij}^{(m)}\right)$$

can be viewed as being sufficiently close to zero.

4.2.2.2 Gravity Model

The growth factor models do not take into account the relative travel costs associated with alternative destinations. These cost can change in time because of new roadway construction, or other changes in the general area, including mass transit schedules and availability. The Gravity Model, (see Voorhees, 1955), is one that tries to take into account the degree of attraction between two zones. It states that the number of trips between two zones is proportional to the total number of trips originating at the origin zone i, the number of potential trip ends at the terminating zone j, and a cost function F_{ij}, also called a "friction" between the two zones. If the zones are near each other, F_{ij} is large, and if they are far from each other, it is relatively small. For some observed, base period, travel patterns, the model states that

$$t_{ij} = t_{Ei}\left(\frac{A_j F_{ij}}{\sum_{n=1}^{N} A_n F_{in}}\right), \tag{13}$$

where A_j is a measure of attractions, and useful trip ends, characterizing node j. Equation 13 states that trips are apportioned according to the relative attractiveness of various possible end nodes. The sum in the denominator guarantees that

$$\sum_{j=1}^{N} t_{ij} = t_{Ei} \tag{14}$$

namely, that the sum of the trips from zone i to all other zones is equal to the exogenously obtained estimate of this sum. The measure of attraction A_j may be defined for a particular class of trips, e.g. travel to work, or it may lump together all attractions. In the latter case, if $t_{ij} = t_{ji}$, then $A_j = t_{Ei}$ may be an appropriate value.

The next challenge is to find an appropriate function F_{ij} which can be applied to zones i and j. A good first approximation is to assume that $F_{ij} = F(d_{ij})$, where d_{ij} is the travel time between the two zones, and consequently an inverse measure of attractiveness. Assuming an empirical function $F(d)$, we can calibrate the model by minimizing the sum of the squared deviations, $(t_{ij} - t_{Eij})^2$ between the model and actual measurements. If such a function is found, it is generally assumed that its form will remain valid for at least an appreciable length of time, and thus future trip distribution can be obtained as

$$T_{ij} = T_{Ei}\left(\frac{A_j F_{ij}}{\sum_{n=1}^{N} A_n F(d_{in})}\right) \tag{15}$$

where A_j, T_{Ei}, and d_{ij} are the best estimates of future attractions, trip generation, and travel times. It may be noted here that the value of F_{ij} may be specifically chosen in order to take into account such things as a major barrier between zones, or a costly toll facility. For example, we may set $F_{ij} = F(d_{ij})R_{ij}$, where R_{ij} is a correction factor accounting for the special impediments to travel between i and j, and is equal to 1 if no such impediments are present. A popular choice for $F(d_{ij})$ is

$$F(d_{ij}) = \frac{1}{d_{ij}^{c}} \quad , \tag{16}$$

where the exponent c is a constant fitted to available data. Observation of Eq. 16 reveals the origin of the term "gravity model", because of its resemblance with the Newtonian law of gravitational attraction of two masses, which corresponds to $c = 2$.

4.2.2.3 Intervening Opportunity Model

The Intervening Opportunity Model, as discussed by Ruiter (1967), is based on the assumption that, for every trip originating at a particular zone, all appropriate possible destinations are ordered on the basis of the travel time from the zone, in order of increasing travel time. It is also assumed that there is a constant probability p that a considered site is accepted as the trip end. The remainder of the formulation is analogous to the one used for the intervening opportunity model for land use, Section 4.2.1.

Let $M(d)$ be the number of eligible destinations nearer than d to the trip origin, where d is commonly viewed as the travel time. Assuming that the number of destinations is large, we can treat $M(d)$ as a continuous variable. Then, the probability of rejecting the nearest M sites, denoted by $P(M)$, can be deduced from

$$P(M + dM) = P(M)(1 - p\,dM) \tag{17}$$

which has the solution

$$P(M) = e^{-pM} \tag{18}$$

Accordingly, the value of t_{ij} is given by

$$t_{ij} = t_{Ei}\left(p_{j_1} - p_{j_2}\right) \tag{19}$$

where

p_{j1} = probability of not accepting any destination nearer than zone j,
p_{j2} = probability of not accepting any destination in zone j or nearer than zone j.

Therefore,

$$t_{ij} = t_{Ei}\{\exp(-pM_j) - \exp[-p(M_j + m_j)]\} \tag{20}$$

where M_j is the number of destinations nearer to the trip origin than the destinations in zone j, and m_j is the number of destinations in zone j. Equation 20 may now be used to calibrate the model according to available data, by finding the best possible value of p that fits the data. After calibration, Eq. 20 may be used to estimate T_{ij} in terms of T_{Ei}.

4.2.2.4 Minimum Entropy Formulation

The gravity model discussed earlier is a proportional model, in the sense that the number of trips from zone i to zone j is proportional to the total number of trips, t_{Ei}, from zone i, and to the number of attractions, A_j, in zone j. The model may be simplified by retaining the proportionality feature but ignoring "friction", and therefore setting $F_{ij} = 1$ for all pairs of i and j, in which case Eq. 13 becomes

$$t_{ij} = \frac{t_{Ei} A_j}{\sum_{n=1}^{N} A_n} \tag{21}$$

Over a 24 hour period, we may assume that all trips are associated with a return trip, therefore $t_{ij} = t_{ji}$, and $A_j = t_{Ej}$ = the number of trips observed or estimated as originating from zone j. Then, the conditions satisfied by this simplified model are

$$\begin{aligned} \sum_{j=1}^{N} t_{ij} &= t_{Ei} , \\ \sum_{i=1}^{N} t_{ij} &= A_j = t_{Ei} , \\ \sum_{i=1}^{N} \sum_{j=1}^{N} t_{ij} &= \sum_{j=1}^{N} t_{Ej} = T , \end{aligned} \tag{22}$$

where T is the total number of trips in the network. The conditions in Eqs. 22 express the *conservation laws*, namely, that every trip has an origin and a destination, and the total number of trips in the area is equal to the sum of all zonal trips.

It may now be shown that the same Eq. 21 can be obtained by minimizing the function

$$S = \sum_{i=1}^{N}\sum_{j=1}^{N} t_{ij} \log t_{ij} \tag{23}$$

subject to the basic conditions of Eqs. 22. At this point we have an explanation of the term *entropy* used in the label of this model. Because the function S looks very much like an entropy function, as used in statistical thermodynamics, the model was labeled as the entropy model, (see Potts and Oliver, 1972, pp. 121-135).

The minimization of S in Eq. 23 subject to the conditions in Eqs. 22 may be done through the use of Lagrange multipliers, involving minimizing the composite function

$$F = S - \sum_{i=1}^{N} a_i \left(t_{Ei} - \sum_{j=1}^{N} t_{ij} \right) - \sum_{j=1}^{N} b_j \left(t_{Ej} - \sum_{i=1}^{N} t_{ij} \right) - c\left(T - \sum_{i=1}^{N}\sum_{j=1}^{N} t_{ij} \right), \tag{24}$$

where a_i, b_j, and c are undetermined multipliers. To minimize F, we take the derivatives with respect to t_{ij}, a_i, b_j, and c and set them equal to zero, thus obtaining

$$\frac{\partial F}{\partial t_{ij}} = \log t_{ij} + 1 - a_i - b_j - c = 0 \tag{25}$$

plus the constraints expressed by Eqs. 22. The solution of Eq. 25 is

$$t_{ij} = \exp(a_i + b_j + c - 1) \tag{26}$$

We now enter t_{ij}, as expressed in Eq. 26, into Eq. 22, then multiply the first two relations of Eqs. 22 and divide the result by the third expression, obtaining

$$t_{ij} = \exp(a_i + b_j + c - 1) = \frac{t_{Ei} t_{Ej}}{T} = \frac{t_{Ei} A_j}{\sum_{j=1}^{N} A_j} \quad . \tag{27}$$

Since the result shown in Eq. 27 is the same as in Eq. 21, the entropy formulation is shown to be equivalent to the simplified gravity model, under the assumptions used for the various parameters. More complex versions of the entropy model may be obtained, for example, by taking into account explicitly the relative travel times between zones. Extensions to the above model have been given by Wilson (1967). Although the use of the entropy concept is not universally endorsed, (see for example Beckman and Golob, 1972), it does present a capability of handling models of increasing complexity in a rational manner.

4.3 TIME-INDEPENDENT TRAFFIC ASSIGNMENT

Having the origin-destination matrix for a network, the next and most important question is how traffic proceeds from origin to destination. This is a matter of *route choice* by an individual driver. Today, and in the foreseeable future, such a choice is strictly in the hands of the driver. An attempt to describe the driver's choice was first made by Wardrop (1952), who proposed the following two principles:

1. The journey times on all routes actually used are equal and less than those which would be experienced by a single vehicle on any unused route,
2. The average journey time (for all users) is minimum.

It is assumed that principle #1 is followed by individual users in selecting their routes, and for this reason it is known as leading to the "user-optimal" solution to the traffic assignment problem. Principle #2 is known as leading to the "system-optimal" solution. It turns out that no matter how well the user-optimal solution is implemented, it does not also guarantee a system-optimal solution, because in general the two principles are not satisfied at the same time. This incongruence of the two solutions, although not very paradoxical, is a legitimate feature of the "Braess Paradox" mentioned in Chapter 1 of this book.

An obvious application of Wardrop's first principle is to assume that a driver tries to follow the shortest route to a destination. Shortest route computations are routine in graph theory, and they have been applied to transportation networks. One of the special nuances in transportation networks is that the travel time along a link is not constant but depends on the volume of traffic along the link. Phenomenological relationships between speed and density have been applied to account for this variability of transit time along network links. Other models used to describe human behavior and determine traffic assignment characteristics, take into account a set of attributes of different route choices, which may include financial as well as physical characteristics of networks.

We shall now discuss traffic assignment models proposed over the years, including both deterministic models and stochastic models. We shall first consider the case of time-independent traffic assignment, in which the traffic demand is assumed to be essentially constant. We shall start with a discussion of deterministic traffic assignment models, in which the travel time or cost is assumed to be precisely known by the users of the system. We shall then go on to stochastic models, in which some uncertainty is allowed concerning the properties of the network. In particular, we shall discuss the *Discrete Choice* models, which are a natural sequel to the preceding discussion of traffic generation, and to some extent integrate traffic generation and traffic assignment. We shall then proceed to the discussion of both deterministic and stochastic models of the time-dependent variety, known as *Dynamic Traffic Assignment*. We shall conclude with a discussion of the effect of congestion on traffic assignment computations.

4.3.1 Deterministic Traffic Assignment Models

The development of Traffic Assignment models has been based on the application of the two Wardrop principles, and particularly the first one. This is natural, because by and large decision regarding travel choices are still in the hands of individual drivers, who are likely to apply Wardrop's first principle, if they apply anyone at all. Wardrop's second principle would be implemented only through the intervention of traffic management authorities interested in minimizing the aggregate cost of travel to all users. Such an intervention has not been forthcoming as yet, notwithstanding the efforts at developing Intelligent Transportation Systems (ITS), in which interaction with drivers may ultimately influence their travel choices.

It should be pointed out at the start, that all models proposed for handling driver choices from among alternatives implicitly assume superior wisdom and knowledge on the part of a driver, plus willingness to make drastic changes in order to achieve optimality. This, however, is not the case with most drivers. By and large, drivers may be knowledgeable about one or two alternate routes to a destination, and may be willing to switch to one or the other if there is convincing evidence that the alternative choice leads to a congested domain of the network. Nevertheless, we shall discuss the ambitious traffic assignment algorithms, with the thought that increasing deployment of ITS services is likely to increase the knowledge, and decision-making ability of drivers.

4.3.1.1 Cheapest Routes

There is a plethora of algorithms for determining the cheapest routes on a network, with an extensive bibliography provided by Murchand (1969). All of them are essentially based on the following premise: If a cheapest route from node n_1 to node passes through n_i, then the portion of the route from n_1 to n_i is the cheapest route from n_1 to n_i, and similarly the portion of the route from n_i to n_r is the cheapest route from n_i to n_r. The reader will recognize this a basic statement of a *dynamic program* as discussed by Bellman (1958). Accordingly, the cheapest routes from node n_1 to node n_r of a network [N ; L] with positive link costs $c(n_i, n_j)$ are the unique solutions of the functional equations

$$f(n_1) = 0 ,$$
$$f(n_r) = \min_{n_i \neq n_r}[f(n_i) + c(n_i, n_r)], \quad n_r \neq n_1 \quad . \tag{28}$$

The various cheapest route algorithms are, in effect, procedures for solving Eqs. 28. We start with the *Tree-Building Algorithms.*

4.3.1.2 Tree-building Algorithms

This algorithm consists of fanning out from the home node (origin) to all other nodes in increasing order of their costs from the origin, as described by

Dijkstra (1959). The nodes are successively labeled with two numbers, one describing the *predecessor* node on a cheapest route from the origin, and the other the cost of the route up to the node in question. Initially, the origin is labeled (0,0), and all the other nodes are temporarily labeled $(0,\infty)$. A node is permanently labeled, when the cheapest route to it is determined. The following notation is used in the sequel:

- The graph of the network is a *directed graph* $[N; L]$, consisting of N "nodes", which are denoted by i or n_i, ($i = 1,2,\ldots,n$), and L "links" denoted by (i, j) or (n_i, n_j).
- A *partial graph* of a directed graph $[N; L]$ is a directed graph $[N; L']$ with L' a subset of L, i.e. $L' \subseteq L$.
- A *subgraph* of $[N; L]$ is a directed graph $[N'; L']$ with $N' \subseteq N$, and where L' is the set of all links of L which join the nodes of N'.
- A *bipartite graph* is one in which the set of N nodes is partitioned into two complementary sets $X, \overline{X}$, i.e.

$$X \cup \overline{X} = N, \qquad X \cap \overline{X} = \varnothing = \text{empty set.}$$

The algorithm proceeds as follows. The last node permanently labeled is considered as a *predecessor* node, and all nodes to which links from this node are directed are looked at as possible new predecessor nodes. If the sum of the cost to the current predecessor node plus the link cost to a downstream node is lesser than the tentative cost label to that node, then this new cost becomes a new tentative cost label for the downstream node, and all other tentative labels are left unchanged. After all the possible downstream nodes are investigated, the one corresponding to the minimum new tentative cost is chosen as the new predecessor node, its tentative cost becomes permanent, and the process is repeated until the permanent labeling is completed.

It is convenient to use a duplicate notation for all nodes, by denoting them as $1,2,\ldots,i,\ldots,n$, and also $n_1, n_2, \cdots, n_k, \cdots$, as they are successively selected at steps $1,2,\ldots,k,\ldots$ of the algorithm. For simplicity, it will be assumed that the network $[N; L]$, (with N nodes and L links), is connected and undirected and that the link costs are $c(n_i, n_j) > 0$. At the kth step, nodes n_j are tentatively labeled with two numbers, one being $P^{(k)}(n_j)$, the current predecessor node, and the other $C^{(k)}(n_1, n_j)$, the cost of the currently known cheapest route from the origin n_1 to node n_j. When the algorithm is concluded, all nodes have permanent labels $P^*(n_j)$ and $C^*(n_1, n_j)$, representing the predecessor node and the cost of the cheapest path, respectively. The numerical procedures for the algorithm are as follows:

Step $k = 1$:

$$\begin{aligned} & X_1 = \{n_1\}, \\ P^*(n_1) = 0 \quad & C^*(n_1, n_1) = 0, \\ P^{(1)}(n_j) = 0 \quad & C^{(1)}(n_1, n_j) = \infty. \end{aligned} \tag{29}$$

Steps $k = 2,3,\ldots,n$:

a) For $n_j \in \overline{X}_{k-1}$, $(n_{k-1}, n_j) \in (X_{k-1}, \overline{X}_{k-1})$, and $C^*(n_1, n_{k-1}) + c(n_{k-1}, n_j) < C^{(k-1)}(n_1, n_j)$,

set

$$\begin{aligned} P^{(k)}(n_j) &= n_{k-1}, \\ C^{(k)}(n_1, n_j) &= C^*(n_1, n_{k-1}) + c(n_{k-1}, n_j). \end{aligned} \tag{30}$$

b) For all other $n_j \in \overline{X}_{k-1}$, define

$$\begin{aligned} P^{(k)}(n_j) &= P^{(k-1)}(n_j), \\ C^{(k)}(n_1, n_j) &= C^{(k-1)}(n_1, n_j). \end{aligned} \tag{31}$$

c) Define n_k by

$$C^{(k)}(n_1, n_k) = \min_{n_j \in \overline{X}_{k-1}} C^{(k)}(n_1, n_j), \tag{32}$$

and define the permanent labels

$$P^*(n_k) = P^{(k)}(n_k) ,$$
$$C^*(n_1, n_k) = C^{(k)}(n_1, n_k) . \tag{33}$$

d) Define

$$X_k = X_{k-1} \cup n_k \quad . \tag{34}$$

Since $C^*(n_1, n_k) \geq C^*(n_1, n_{k-1})$, the above procedure solves Eqs. 28 by permanently labeling all nodes in order of increasing cost from the origin. The cheapest path themselves are determined by tracing back to the origin across successive predecessor nodes. This completes the tree-building algorithm.

In the above discussion, we talked about *cost* associated with a link. This cost is usually strongly tied to the travel time along the link, but additional factors may be considered, such as *turn penalties and prohibitions*. In the end, the decision-making individual is assumed to select the route corresponding to the cheapest route, according to Wardrop's first principle, leading to the *Cheapest Route Assignment* for every individual, but not necessarily to the minimum aggregate delay for all individuals combined. It may be pointed out here that the procedure just outlined assumes a great deal of flexibility, and superior wisdom, on the part of all users of the network, a rather unlikely occurrence. Quite often, users limit themselves to just a few route choices, making the route assignment much simpler , albeit somewhat removed from the optimum. In traffic assignment practice, the most common cheapest route assignment is the *all-or-nothing assignment* , in which link costs are supposed to be constant and flow-independent. Traffic between any origin and destination pair is all assigned to the cheapest route, and none is assigned to any other route. Such an algorithm is justifiable for not too heavy traffic, but it becomes increasingly inaccurate as congestion sets in.

4.3.2 Discrete Choice, Stochastic, Models

The following is a concise overview of the primary *Discrete Choice Models*, (DCM), proposed over the years. Extended discussions of the topic may be found in Ben-Akiva and Lerman (1985) , and in Ben-Akiva and Bierlaire (1999).

Decision Maker. In a DCM, a *Decision-Maker* who is effectively an individual, but may also be a group making a single decision on choice, effectively goes through the following steps:

- Determines the options available,
- Estimates the benefits and costs of these options,
- Applies a decision rule to choose the "best" option.

Alternatives. By necessity, the alternatives presented to decision-makers are limited to those of which they are aware. This limitation introduces an uncertainty in modeling the generation process of choice sets, which has motivated the use of probabilistic models that predict the probability of each feasible choice set being selected from among all possible choice sets.

Attributes. Each alternative is associated with various attributes, which may be shared by other alternatives. An attribute may not be a measurable quantity, but may be any function of available data. Some assumptions of human behavior may dictate the form of such functions. For example, out-of pocket cost of travel may be considered alone as an attribute, but it may be found that a better correlated attribute is the out-of-pocket cost divided by the income of the decision maker.

Decision Rules. A decision-maker implicitly evaluates the attributes of various route choices and determines a choice. Most models of decision-making are based on *utility theory*, which involves the association of each choice with a merit value called *utility*, after which the decision-maker chooses the alternative with the highest utility. For example, if the decision-maker is only interested in minimizing travel time, the utility of the various route choices would be proportional to the inverse of the travel time.

The utility theory assumption regarding decisions by individuals is elegant, but not altogether realistic. Human beings are not generally able to assess attributes reliably, and they often make almost arbitrary decisions which deviate from those that utility theory alone would dictate. For this reason, models involving probabilistic decision rules have been proposed to account for deviations from the deterministic application of utility theory. Some of these models assume deterministic utilities and probabilistic decision rules. Examples are the model of Luce (1959), and the "elimination by aspects" model of Tversky (1972). Other models, used extensively in econometrics as well as travel behavior analysis, and known as *Random Utility Models*, are based on deterministic decision rules and utilities represented by random variables.

4.3.2.1 Random Utility Models

Random utility models assume that the decision maker has a perfect discrimination capability, but the analyst who tries to guess the decision making process has incomplete information, and therefore some uncertainty will exist regarding the decision maker's choices. The utility is modeled as a random variable, reflecting this uncertainty. Specifically, the utility that individual n associates with alternative i in a choice set C_n is assumed to be given by

$$U_{in} = V_{in} + \varepsilon_{in} \,, \tag{35}$$

where V_{in} is the deterministic part of the utility, and ε_{in} is a random term accounting for the uncertainty. It is assumed that the decision maker chooses the alternative with the highest utility, therefore, the probability that the alternative i is chosen from the choice set C_n is

$$\mathrm{P}(i|C_n) = \mathrm{P}\{U_{in} \geq U_{jn} \forall j \in C_n\} = \mathrm{P}\left\{U_{in} = \max_{j \in C_n} U_{jn}\right\}. \tag{36}$$

To make the random utility model operational, we need to introduce certain additional assumptions.

Location and scale parameters. If the values of U_{in} are replaced by ($\mu U_{in} + \alpha$), where μ and α are arbitrary real numbers, Eq. 36 is still valid, as can be seen by the following derivations:

$$\begin{aligned} &\mathrm{P}\{U_{in} \geq U_{jn} \forall j \in C_n\} = \\ &\mathrm{P}\{\mu U_{in} + \alpha \geq \mu U_{jn} + \alpha \forall j \in C_n\} = \\ &\mathrm{P}\{U_{in} - U_{jn} \geq 0 \forall j \in C_n\}. \end{aligned} \tag{37}$$

It is seen that only the signs of the differences between utilities determine the decision outcome, and not the utilities themselves. The constants μ and α are known as the *scale* and *location* parameters, respectively, and they are chosen for various objectives. The scale factor is usually chosen for an appropriate normalization of one of the variances of the random terms, while the location parameter α is usually set to zero.

Alternative Specific Constants. Without undue restriction, the mean of the random terms can be assumed to be equal to any convenient value c. If we denote the mean of the error term of alternative i by $m_i = \mathrm{E}\,[\varepsilon_{in}]$, we can define a new random variable $e_{in} = \varepsilon_{in} - m_i + c$, such that $\mathrm{E}\,[\varepsilon_{in}] = c$. This leads to a model in which the deterministic parts of the utilitiesare $V_{in} + m_i$ and the random terms are e_{in} with mean c, satisfying the relationship

$$\mathrm{P}\{U_{in} \geq U_{jn} \forall j \in C_n\} = \mathrm{P}\{V_{in} + m_i + e_{in} \geq V_{jn} + m_j + e_{jn} \forall j \in C_n\}. \quad (38)$$

The terms m_i are known as *Alternative Specific Constants* (ASC), and they capture the means of the random errors. Inclusion of ASCs in the deterministic part of the utility functions allows, without loss of generality, for the error terms of random utility models to have a constant mean c. The choice of the ASCs is generally dictated by reasons of computational convenience, although it may affect the estimation process of the decision making, as shown by Bierlaire et al (1997).

Utility Value. The utility value is divided into the deterministic part, and the stochastic part.

The deterministic term V_{in} for each alternative is a function of the attributes of the alternative as well as the characteristics of the decision-maker, given by

$$V_{in} = V(z_{in}, S_n) \quad (39)$$

where z_{in} is the vector of attributes for attribute i as perceived by the individual n, and S_n is the vector of characteristics of the individual. A new vector of attributes is now defined from both z_{in} and S_n by using an appropriate vector valued function h, according to

$$x_{in} = h(z_{in}, S_n) \tag{40}$$

where the function h is generally assumed to be continuous and monotonic in z_{in}, and its form is chosen to give the best representation in a specific application. Equation 39 now becomes

$$V_{in} = V(x_{in}), \tag{41}$$

and if the utility specification is linear in the parameters,

$$V_{in} = \sum_k \beta_k x_{ink} , \tag{42}$$

with the utility fully specified by the vector of parameters β.

The random part of the utility can have any of several form, and we shall describe here the ones used in two leading families of utility functions, the *Logit* family, and the *Probit* family.

4.3.2.2 The Logit Family

The Logit (Logistic Probability Unit) model was first introduced in the context of binary choice, where the logistic distribution is used. When generalized to more than two choices of alternatives, it is known as the *Multinomial Logit Model*. It is based on the assumption that the error terms are independent and Gumbel-distributed (Gumbel, 1958) , that is, for all *i, n* it is described by

$$\begin{aligned} F(\varepsilon) &= \exp\{-e^{-\mu(\varepsilon-\eta)}\},\ \mu > 0 , \\ f(\varepsilon) &= \mu e^{-\mu(\varepsilon-\eta)} , \end{aligned} \tag{43}$$

where η is a location parameter, and μ is a strictly positive scale parameter. The mean of the distribution of Eq. 36 is $(\eta + \gamma / \mu)$, where

$$\gamma = \lim_{k \to \infty} \sum_{i=1}^{k} \left(\frac{1}{i} - \ln k \right) \cong 0.5772$$

is the *Euler constant.* The variance of the distribution is $\pi^2 / 6\mu^2$, and the probability that individual n chooses alternative i from the choice set C_n is given by

$$\mathrm{P}(i|C_n) = \frac{e^{\mu V_{in}}}{\sum_{j \in C_n} e^{\mu V_{jn}}} \quad . \tag{44}$$

An implicit property of the Multinomial Logit Model is a basic lack of correlation among alternatives, which sometimes leads to strange, counter-intuitive results. An extension to the Multinomial Logit Model, designed to capture some of these correlations, was first proposed by Ben-Akiva (1974), and is known as the *Nested Logit Model* . It involves partitioning the choice set C_n into M "*nests*", (C_{mn} , $m \in M$), with the utility of each alternative comprising two terms, one specific to the alternative and the other associated with the nest. The Nested Logit Model corrects some of the deficiencies of the Multinomial Logit Model by capturing choice problems in which alternatives within each nest are correlated. No correlation across nests is allowed by the Nested Logit Model, which therefore cannot be used when alternatives cannot be partitioned in well separated nests of cross-correlated alternatives.

An extension of the Nested Logit Model is the *Cross-nested Logit Model*, in which an alternative may belong to more than one nest. It was first proposed by McFadden (1978) as a special case of the *Generalized Extreme Value Model* (GEV), which includes the Multinomial Logit and the Nested Logit models. In the GEV model, the probability of choosing alternative i within C_n is

$$P(i|C_n)=\frac{e^{V_{in}}\dfrac{\partial G}{\partial e^{V_{in}}}\left(e^{V_{1n}},\cdots,e^{V_{jn}}\right)}{\mu G\left(e^{V_{1n}},\cdots,e^{V_{Jn}}\right)} \tag{45}$$

where G is a non-negative differentiable function, with the following form for different Logit Models, (J_n = number of alternatives in C_n):

1. Logit model: $G(x)=\sum_{i=1} x_i$,
2. Nested Logit Model:
3. Cross-nested Logit Model: $\sum_{m=1}\left(\sum_{j\in C_n} \quad {}_{jm} \quad {}_{j} \right)$

4.3.2.3 Multinomial Probit Model

This model has been named after the "Probability Unit" (Probit) which characterizes it. It is derived from the assumption that the error terms of the utility functions are normally distributed, and captures explicitly the correlation among all alternatives.

We define the (J_n x 1) vectors U_n, V_n and ε_n, where U_n are the utility functions $U_n = V_n + \varepsilon_n$. The vector of error terms $\varepsilon_n = (\varepsilon_{1n}, \varepsilon_{2n}, \cdots, \varepsilon_{Jn})^T$ is multivariate, normal-distributed, with a vector of means 0 and a J_n x J_n variance-covariance matrix Σ_n . The probability that the individual n chooses alternative i from the choice set C_n is given by

$$P(i|C_n) = P(U_{jn} - U_{in} \le 0 \,\forall\, j \in C_n) \ . \tag{46}$$

We now define the $\{(J_n - 1)$ x $J_n\}$ matrix Δ_i according to

$$\Delta_i U_n = \left[U_{1n} - U_{in}, \cdots, U_{(i-1)n} - U_{in}, U_{(i+1)n} - U_{in}, \cdots, U_{J_n n} - U_{in}\right]^T , \tag{47}$$

which turns out to satisfy the relationship

$$\Delta_i U_n \approx \mathrm{N}\left(\Delta_i V_n , \Delta_i \Sigma_n \Delta_i^T\right) \ . \tag{48}$$

The density function is now given by

$$f_i(x) = \lambda \exp\left\{-\frac{1}{2}(x - \Delta_i V_n)^T \left(\Delta_i \Sigma_n \Delta_i^T\right)^{-1} (x - \Delta_i V_n)\right\} , \tag{49}$$

where

$$\lambda = (2\pi)^{\frac{J_n - 1}{2}} \left|\Delta_i \Sigma_n \Delta_i^T\right|^{-1/2} , \tag{50}$$

and

$$\mathrm{P}(i|C_n) = \mathrm{P}(\Delta_i U_n \le 0) = \int_{-\infty}^{0} \ldots \int_{-\infty}^{0} f_i(x) dx_1 \cdots dx_{i-1} dx_{i+1} \cdots dx_{Jn} \ . \tag{51}$$

The matrix Δ_i has a form in which the ith column contains -1 across all rows. If the ith column is removed, the remaining $(J_n - 1) \mathrm{x} (J_n - 1)$ matrix is the identity matrix.

It should be pointed out that the multi-fold integral of Eq. 51 becomes intractable even for a relatively low number of alternatives, and the number of unknown parameters in the variance-covariance matrix grows with the square of the number of alternatives, making computations rather cumbersome. A detailed discussion of multinomial probit models can be found in McFadden (1989) . An additional variation on the theme of Multinomial Probit has been the *Hybrid Logit Model* proposed by Ben-Akiva and Bolduc (1996), which is intended to bridge the gap between Logit and Probit models by combining the advantages of both.

4.3.3 Global Network Optimization Models

So far, we have considered networks subject to Wardrop's first principle, namely, networks in which route decisions are in the hands of individual drivers. We shall now discuss the implications of Wardrop's second principle, namely, the case when intervention of a higher authority may force individuals to select routes which optimize the operation of the network for the aggregate of users. In increasing level of complexity, we can define the following types of networks, which are to be optimized:

1) Networks with fixed link costs, and no link capacity restraints.
2) Capacitated networks, with fixed link costs.
3) Networks with flow-dependent costs.

4.3.3.1 Fixed Link Costs, no Capacity Restraints

Let us first dispense with the trivial case corresponding to fixed link costs, and no capacity restraints. This is the case in which it is assumed that the performance of a link is not degraded by the addition of traffic, and such addition is unconstrained. If, for example, the cost is tied to travel time, the assumption here is that this travel time does not change appreciably with increasing traffic demand. This is clearly a case of a network with rather light traffic, operating in the range of densities for which speed is almost constant, and flow is a linear function of the density. In this case, the two Wardrop principles are simultaneously satisfied. Optimization of individual choices results in a global optimum for the operation of the network, since the presence of any individual on a link does not affect the performance of that link. Even discrete choice models, like the Logit and Probit models, would be associated with a global optimum after individual optimization, as long as the traffic characteristics affecting individual choices, including uncertainties, are not influenced by traffic demand and traffic densities.

It will be useful for the discussion in this section, and others to follow, to define certain variables and illustrate them by an example. Consider the network depicted in Fig. 7. It is assumed that a number of OD pairs have

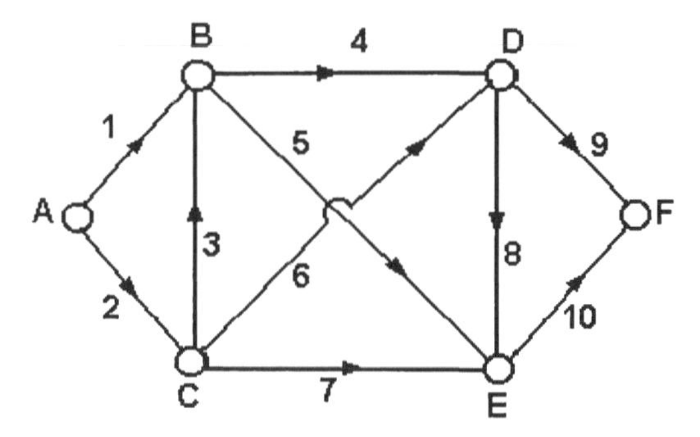

Figure 7. Traffic Assignment in a Network

been defined for this network, and that the link costs c_j for the links j = 1,2,...,10 are given. For example, let us assume that two OD pairs exist, A-F, and B-F. For each OD pair, there exist a number of link paths. In the example of Fig. 7, the possible paths, nine altogether, are:

For A-F : 1,4,9 2,7,10 1,5,10 2,6,9 1,4,8,10 2,3,4,9
For B-E : 4,9 5,10 4,8,10

We denote the paths by P_k, k = 1,2,...,K, where K is the total number of paths which can serve all OD pairs. We can define the *link incidence matrix*, which associates each link with any of the above paths, by assigning the value 1 if the link is part of the path, and 0 if it is not. For example, if the A-F paths are numbered from 1 to 6, and the B-F paths from 7 to 9, the link incidence matrix $E_{kj}, (k = 1,2,\cdots,K), (j = 1,2,\cdots,J)$ is given by Eq. 52 . If we denote the link costs by c_j, (j = 1,2,...,J) we can compute a vector of path costs, $C_k, (k = 1,2,\cdots,K)$ according to Eq. 53. It is easy then to identify the cheapest route corresponding to any specific OD pair. This cheapest route will be the one chosen by a knowledgeable individual associated with the OD pair. It will also be the path chosen for the purpose of minimizing the network cost for the aggregate of users, since the link costs are not affected by the volume of traffic associated with them.

$$E_{kj} = \begin{bmatrix} 1 & 0 & 0 & 1 & 0 & 0 & 0 & 0 & 1 & 0 \\ 0 & 1 & 0 & 0 & 0 & 0 & 1 & 0 & 0 & 1 \\ 1 & 0 & 0 & 0 & 1 & 0 & 0 & 0 & 0 & 1 \\ 0 & 1 & 0 & 0 & 0 & 1 & 0 & 0 & 1 & 0 \\ 1 & 0 & 0 & 1 & 0 & 0 & 0 & 1 & 0 & 1 \\ 0 & 1 & 1 & 1 & 0 & 0 & 0 & 0 & 1 & 0 \\ 0 & 0 & 0 & 1 & 0 & 0 & 0 & 0 & 1 & 0 \\ 0 & 0 & 0 & 0 & 1 & 0 & 0 & 0 & 0 & 1 \\ 0 & 0 & 0 & 1 & 0 & 0 & 0 & 1 & 0 & 1 \end{bmatrix} \tag{52}$$

$$C_k = \sum_{j=1}^{J} E_{kj} c_j \ , \tag{53}$$

As a numerical example, let us assume that the link costs are

$$c_j = (3, 4, 5, 5, 7, 8, 6, 6, 4, 3).$$

The cost vector would then be

$$C_k = (\ 12,\ 13,\ 13,\ 16,\ 17,\ 18,\ 9,\ 10,\ 14) \ ,$$

where the first six components correspond to the OD pair A-F, and the last three to the pair B-F. It can be seen that the optimum assignment corresponds to the choice of path 1,4,9 for the A-F pair, and 4,9 for the B-F pair. The solution was thus obtained by complete enumeration of all choices, but it can also be formulated as a Linear Programming problem, as discussed by Potts and Oliver (1972).

4.3.3.2 Capacitated Networks with Fixed Link Costs

Let us now assume that each link j is *capacitated*, namely, it can handle traffic volumes only up to a maximum, Q_j. Minimizing the network cost must now take into account explicitly the volumes $G_l, (l = 1,2,\cdots,L)$ associated with the L OD pairs, where $L = 2$ in the above example. If these volumes cannot be fully accommodated by allocation to the cheapest route for each OD pair, we must partition them among all possible paths. Let us assume that the volume of the lth pair is partitioned among the K paths, in proportion to the variables α_{li}, $(i = 1,2,\cdots,K)$, where all the paths are included in K, whether or not they correspond to the lth pair. If they do not, the corresponding α_{li} is automatically zero.

We now compute a flow matrix F_{kj} for all paths and links, according to

$$F_{kj} = \sum_{l=1}^{L} \alpha_{lk} E_{kj} \ . \tag{54}$$

We have the following relationships:

$$\begin{aligned} &\sum_{k=1}^{K} \alpha_{lk} = 1 \,, \qquad l = 1,2,\cdots,L \,, \\ &\sum_{k=1}^{K} F_{kj} \le Q_j \ . \end{aligned} \tag{55}$$

Minimizing the aggregate cost to the users is now translated into the objective

$$\min C = \sum_{j=1}^{L} c_j \sum_{k=1}^{K} F_{kj} \,, \tag{56}$$

subject to the constraints expressed by Eqs. 54 and 55.

4.3.3.2.1 Capacitated Networks with Flow-dependent Costs

The optimization problem can become quite complex if the link costs are complicated functions of the link flows. In many cases, traffic theorists have assumed linear relationships between flows and costs, of the type

$$c_j = A + Bf_j \ , \tag{57}$$

where A and B are constants, and f_j is the flow along the link j. The cost functions implied by Eq. 57 are rather crude, and the range of flows over which they may give a reasonably good approximation has to be carefully selected. However, they simplify considerably the optimization problem, which again becomes a Linear Programming problem, whose variables are the flow partitioning variables α_{li}.

As an illustration, let us consider the example of the network of Fig. 7, and the associated link and OD parameters. We shall also assume that the OD demands G_l, and link capacities Q_j, are

$$G_j = \{27, 20\}$$
$$Q_j = \{26, 22, 19, 20, 24, 21, 25, 23, 24, 25\}$$

expressed in units of vehicles per minute. For the purpose of obtaining the minimum cost assignment, we shall also assume linear cost functions for each link, of the type $c_j = A_j f_j$, with

$$A_j = (10, 9, 11, 8, 10, 10, 9, 9, 10, 8) \ .$$

Nine weighting parameters for the nine possible paths are to be determined, namely,

For OD pair A-F: $\alpha_1, \cdots, \alpha_6$,

For OD pair B-F: $\alpha_7, \alpha_8\, \alpha_9$.

The matrix F_{kj} now becomes

$$F_{kj} = \begin{bmatrix} \alpha_1 & 0 & 0 & \alpha_1 & 0 & 0 & 0 & 0 & \alpha_1 & 0 \\ 0 & \alpha_2 & 0 & 0 & 0 & 0 & \alpha_2 & 0 & 0 & \alpha_2 \\ \alpha_3 & 0 & 0 & 0 & \alpha_3 & 0 & 0 & 0 & 0 & \alpha_3 \\ 0 & \alpha_4 & 0 & 0 & 0 & \alpha_4 & 0 & 0 & \alpha_4 & 0 \\ \alpha_5 & 0 & 0 & \alpha_5 & 0 & 0 & 0 & \alpha_5 & 0 & \alpha_5 \\ 0 & \alpha_6 & \alpha_6 & \alpha_6 & 0 & 0 & 0 & 0 & \alpha_6 & 0 \\ 0 & 0 & 0 & \alpha_7 & 0 & 0 & 0 & 0 & \alpha_7 & 0 \\ 0 & 0 & 0 & 0 & \alpha_8 & 0 & 0 & 0 & 0 & \alpha_8 \\ 0 & 0 & 0 & \alpha_9 & 0 & 0 & 0 & \alpha_9 & 0 & \alpha_9 \end{bmatrix} ,$$

and the problem is to minimize the network cost function

$$C = 686\alpha_1 + 646\alpha_2 + 700\alpha_3 + 686\alpha_4 + 826\alpha_5 + \\ 956\alpha_6 + 416\alpha_7 + 430\alpha_8 + 576\alpha_9 \quad ,$$

subject to the constraints

$$\sum_{i=1}^{6} \alpha_i = \sum_{j=7}^{9} \alpha_j = 1 , \qquad 27(a_1 + \alpha_3 + \alpha_5) \le 26 ,$$
$$27(\alpha_2 + \alpha_4 + \alpha_6) \le 22, \qquad 27\alpha_6 \le 19,$$
$$27(\alpha_1 + \alpha_5 + \alpha_6) + 20(\alpha_7 + \alpha_9) \le 23 , \qquad 27\alpha_3 + 20\alpha_8 \le 24 ,$$
$$27\alpha_4 \le 21 , \qquad 27\alpha_2 \le 25 , \qquad 27\alpha_5 + 20\alpha_9 \le 23 ,$$
$$27(\alpha_1 + \alpha_4 + \alpha_6) + 20\alpha_7 \le 24 , \quad 27(\alpha_2 + \alpha_3 + \alpha_5) + 20(\alpha_8 + \alpha_9) \le 28 .$$

The problem statement will be recognized as one of a Linear Programming nature. Its solution is straightforward, and corresponds to the following values of a_j, $j = (1,2,\cdots,9)$:

$$\alpha_1 = 0,\ \alpha_2 = 19/27\ ,\ \alpha_3 = 5/27\ ,\ \alpha_4 = 3/27\ ,\ \alpha_5 = 0,$$
$$\alpha_6 = 0\ ,\ \alpha_7 = 1\ ,\ \alpha_8 = 0,\ \alpha_9 = 0 \quad ,$$

which is translated into the following link flows for the ten links:

$$f_1 = 5\ ,\ f_2 = 22\ ,\ f_3 = 0\ ,\ f_4 = 20\ ,\ f_5 = 5\ ,$$
$$f_6 = 3\ ,\ f_7 = 19\ ,\ f_8 = 0\ ,\ f_9 = 23\ ,\ f_{10} = 24 \quad .$$

It may be observed here that before any attempt at optimization, it is helpful to find out if the network is capable of handling the specified traffic. This can be determined by resorting to *minimal cut* arguments, familiar in graph theory. A minimal cut intersects links downstream from a source or a group of sources, and the summed up capacities of these links is the smallest among those produced by all possible cuts with the same upstream sources. The total throughput of the minimal cut must exceed the sum of the upstream sources if the network is to be able to handle the demands. In the numerical example given, the minimal cut for source A alone is across links 1 and 2, with throughput 48 > 27, (Fig. 7). The minimal cut for both sources A and B is across links 9 and 10, with throughput 49 > 47, so the network capacity is adequate for handling the demands.

4.4 DYNAMIC TRAFFIC ASSIGNMENT

Everything we have covered so far concerning traffic assignment deals with steady-state traffic, unchanging with time. Given the fact that in most urban areas traffic changes substantially with time, particularly during the rush period, it behooves us to ask how such time-dependent traffic should be assigned to different routes in the network. This question has led to the development of *Dynamic Traffic Assignment* methodologies, in which the time-dependent OD demands are explicitly considered as one tries to minimize the travel cost, for a single user or for the aggregate of users. The assignment of paths to different OD pairs will most likely be changing with time in this case. Both, analytical techniques and simulation programs have been used in dealing with dynamic traffic assignment.

One of the earliest contributions on dynamic traffic assignment was that of Yagar (1970, 1971) , who used the simulation approach in order to optimize individual route choice according to Wardrop's first principle. A user was assumed to minimize travel cost for varying traffic demand conditions and queue evolution. The simulation divided time into time slices with constant demand, with queue demand at the end of a given slice added to the new demand. Yagar also contributed a heuristic solution algorithm to the problem. The first mathematical programming approach to the problem was given by Merchant (1974), and by Merchant and Nemhauser (1978a, 1978b). A macroscopic model was formulated for minimizing total transportation cost, in the form of a discrete-time, nonlinear, nonconvex mathematical programming model. Ho (1980), and Carey (1986, 1987) contributed discussions complementing the Merchant-Nemhauser contributions, and in the case of Carey, also reformulating the problem as a well-behaved convex nonlinear problem. Reviews of the topic of Dynamic Traffic Assignment has been provided by Wie (1991), and by Florian and Hearn (1999). Both deterministic and stochastic versions of the dynamic traffic assignment problem have been considered. In what follows, we discuss the basic formulation of a deterministic model.

In dealing with the time-dependent operation of a network, one needs information regarding the time-dependency of the traffic demands over a suitable time horizon. There are two sources for such information:

- Historical data regarding travel demand from different origins to different destinations.
- Real-time traffic data, measuring actual travel demand, as well as detecting traffic incidents which influence the performance of network links.

Consider a traffic network [N ; L] with nodes n, $(n = 1,2,\ldots,N)$, and OD pairs l, $(l = 1,2,\ldots,L)$. There are also k paths in the network, (k = 1.2....,K), whose optimum use is the goal of the optimization problem. A subset K_l of these paths serves the lth OD pair. We consider a time period [0, T] for departures, and a termination time T' at which all traffic is assumed to have been served. Assume that the demand functions for each OD pair are time-dependent, given by $G_l(t)$, $l \in L$, and give rise to path flows $h_k(t), k \in K$, for $t \in [0, T']$. The actual experienced travel time for a path k which carries a flow generated at time t is $s_k(t,h)$. We also assume that the departure rates are continuous over time.

The feasible domain for the dynamic network equilibrium model must satisfy certain flow conservation and non-negativity constraints, namely,

$$\sum_{k \in K_l} k_k(t) = G_l(t),\ l \in L\ ,$$
$$h_k(t) \geq 0,\ k \in K,\ t \in [0, T'] \ . \tag{58}$$

The extension of Wardrop's first (user-optimal) principle may now be stated as follows: *A feasible flow with path flow rates h and path travel times $s(h)$ is a user-optimal solution to the dynamic traffic assignment problem if, at any time t, the actual path travel time for the paths used is the shortest possible among all paths.* This means that we are trying to minimize the expected travel time to individuals at any time $t \in [0, T']$. Therefore, the dynamic traffic assignment is defined by the relationships

$$s_h(t, h^*) - u_l^*(t) = 0 \quad if\ h_k^*(t) > 0\ ,$$
$$s_h(t, h^*) - u_l^*(t) \geq 0 \quad if\ h_k^*(t) = 0\ , \tag{59}$$
$$k \in K_l,\ l \in L.$$

It can be shown that the equilibrium conditions lead to the variational inequality

$$\int_0^T \sum_{l \in L} \sum_{k \in K_l} s_k(t, h^*)[h_k(t) - h_k(t^*)] dt \geq 0\ . \tag{60}$$

It is necessary at this point to state explicitly the order in which traffic demands are served. It is reasonable to assume a First-In-First-Out (FIFO) condition, meaning that a traffic stream cannot overtake another stream that starts earlier. In order to satisfy the FIFO condition when the path costs are additive, (i.e. path costs are the sums of the corresponding link costs), certain conditions must be satisfied by the link cost functions. The FIFO condition may be stated as

$$t' > t'' \Rightarrow t' + s_a(t') > t'' + s_a(t''),\quad a \in A\ , \tag{61}$$

where a denotes a network link from the set of links A. Equation 61 is equivalent to stating that, for a small time increment Δ_t, $(\Delta_t > 0)$,

$$t + s_a(t) < t + \Delta_t + s_a(t + \Delta_t), \tag{62}$$

which, after dividing by Δ_t and taking the limit as $\Delta_t \to 0$, leads to

$$\frac{ds_a(t)}{dt} > -1 . \tag{63}$$

The question then arises as to which link cost functions satisfy the FIFO condition given by Eq. 63. Friesz et al (1993) showed that this condition was satisfied for a linear cost function, and a subsequent study by Xu et al (in press) showed that FIFO conditions would also be satisfied for nonlinear link cost functions which were not "too steep".

Having settled the FIFO issue, we can now look for a solution algorithm for the dynamic network equilibrium model. The algorithm requires a time discretization, and the adaptation of a general method for solving variational inequalities in order to compute a solution of Eq. 60. The solution algorithm may be viewed as comprising two parts. The first is to determine $h_k(t)$, given $s_k(h,t)$. The second, referred to as *the network loading problem*, is to determine $s_k(h,t)$ given the path flow rates $h_k(t)$. A schematic diagram of the necessary process is shown in Fig. 8. It should be pointed out that the solution to the network loading problem is a rather complex task, since it requires the computation of time-varying link flows, travel costs, and path travel costs, by using the time-varying path flow rates $h_k(t)$. A numerical approach to this task has been given by Wu et al (1997).

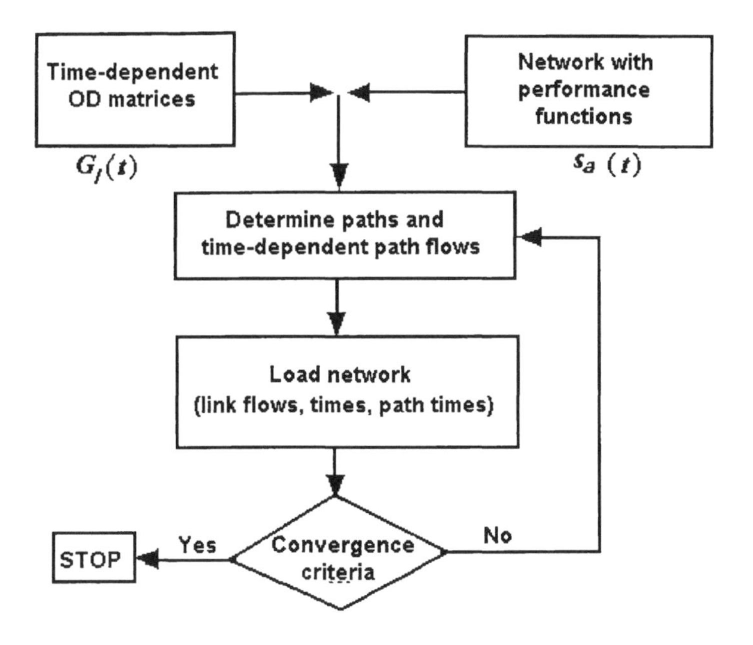

Figure 8. Dynamic Traffic Assignment / Network Loading Problem

It may be pointed out that the preceding solution algorithm does not take into consideration any capacity constraints on the link flows and occupancies. As a result, it is likely to give erroneous results in cases of congestion, when link capacities may be the limiting parameters of any assignment.

There have been additional contributions to the dynamic traffic assignment problem, accounting for uncertainties in the data that influence user decisions. Stochastic models have been proposed, which can account for day-to-day or within-day variations from deterministic determination of network loading, (e.g. Cascetta, 1989). In view of the difficulty in dealing with the mathematics of the dynamic traffic assignment problem in a thorough manner, a great deal of effort has been devoted to simulating dynamic network loading and optimal operation.

4.5 TRAFFIC ASSIGNMENT IN CONGESTED SYSTEMS

In all the preceding sections on traffic assignment, there is a certain element which is characteristic of relatively uncongested systems. It is the fact that the properties and performance of a network link are assumed to depend solely on the traffic volume of that link. Whether one talks about travel cost or something related to that cost, like travel time, the implicit assumption is inherent in all the discussions that these link characteristics are functions of the link traffic volumes alone. Furthermore, they often are assumed to be explicitly represented by phenomenological relationships, such as the flow versus concentration relationships discussed in Chapter 1 of this book.

The above assumptions are increasingly violated as congestion sets in. First of all, it must be realized that the phenomenological flow versus concentration relationship is not like the property of a wire carrying electric current. Traffic links are most often forced to carry higher and higher volumes of traffic because traffic movement is impeded downstream. Insofar as link traffic is relatively uniformly distributed along the link, and moving, one can still use some phenomenological relationships. However, with increasing congestion, this uniformity along the link is violated and the use of such relationships is incorrect. The key observation here is that the link does not stand alone in acquiring its properties, but it is influenced by the downstream section, and similarly it influences the upstream section of a traffic path. The network as a whole becomes a *Store and Forward* system, such as those discussed in Chapter 3 of this book.

As stated already, the basic assumption in the study of traffic assignment is the existence of a so-called "Fundamental diagram" of traffic behavior, such as that discussed in Chapter 1 and shown in Fig. 9. The Fundamental Diagram stipulates that traffic flow, q, is zero at zero traffic density, k, and also at some **Jam Density**, k_j , and it attains a maximum at some intermediate density. The basic premises underlying the Fundamental Diagram are basically correct, although a one-to-one correspondence between flow and concentration is substantially far from what is generally observed in real life. All observations point to a very wide scatter of flow values for any given concentration, particularly at relatively high concentrations. This observation alone should make one cautious in the application of the Fundamental Diagram to describe traffic in a network.

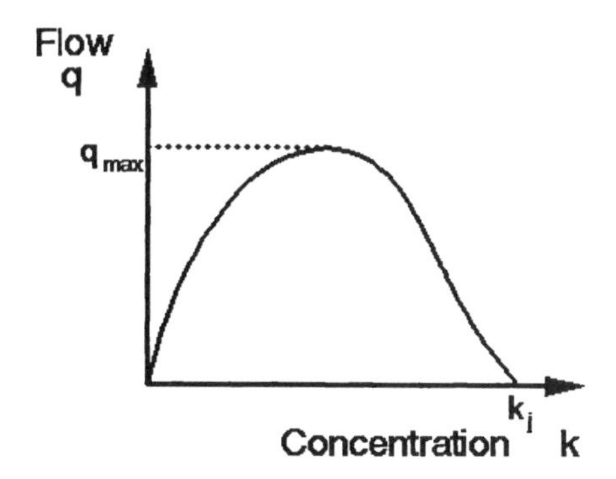

Figure 9. The Fundamental Diagram

However, there is an even greater mistake in such an application. The Fundamental Diagram essentially applies to steady state flow over an infinite highway, and models reasonably well transitions from one steady-state to another over such a highway.

However, an arc of a network is not infinite, and is not described well by a behavior such as that of Fig. 9. A network arc is better described as a traffic facility whose flow may be limited by a **Throughput Limiting Point (TLP)** during periods of congestion. The TLP may be the point of an entrance ramp, the downstream end of an arc, or it may be an intermediate point with a reduced capacity. A collection of TLPs is shown in Figure 10.

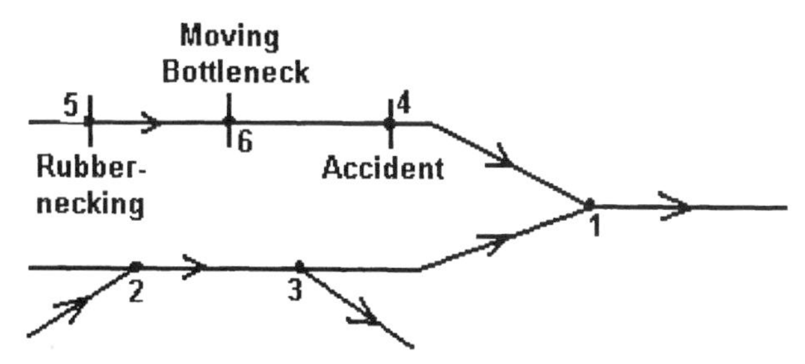

Figure 10. Different types of TLPs.

The nature of the six TLPs shown is as follows:

TLP1: This is the merging of traffic along two arcs into a single downstream arc, where the throughput of the downstream arc is less than the sum of throughputs of the two converging arcs.

TLP2: This is the effect of an entrance ramp feeding into an arc, and thus limiting the throughput allocated to the traffic upstream from the ramp entrance point.

TP3: This is the possible effect of an exit ramp. An exit ramp normally reduces the number of cars in a traffic stream and thus improves the flow. However, much too often, there is a *Spillback* effect caused by a limited capacity of the exit ramp, or an obstruction at the exit ramp such as a traffic light or stop sign at the end of the ramp. As described in a previous paper[2], in such a case a queue is formed along the exit ramp, which spills into the main arc and reduces its throughput.

TLP4: This is a TLP caused by an accident which reduces the throughput of the arc.

TLP5: This is the effect of *Rubber-necking*. Frequently, drivers observing an unusual event, such as an accident on the opposing lanes of a multilane highway, slow down to observe the incident. In doing so, they reduce the throughput of their own arc, thus causing the formation of a TLP

TLP6: This is the effect of a *Moving or Phantom Bottleneck*, described in a previous paper, (Gazis and Herman, 1992). Such a bottleneck may be formed, for example, by a truck climbing along an uphill portion of a highway lacking a slow traffic lane. It becomes a phantom bottleneck when the truck reaches the peak of the uphill portion and resumes normal speed, thus restoring the normal capacity of the arc. However, the effect of the temporary TLP may last for some time after the formation of a TLP6.

TLP1, TLP2, and TLP3 are fixed TLPs, easily identified in a network. TLP4 and TLP5 may be formed at any point in a network, and require good surveillance for their timely discovery. Finally, TLP6 may be an identifiable location, but with a possible intermittent TLP function.

The consequence of the existence of a TLP for a network arc is that exact behavior of the flow versus concentration along the arc is more likely that shown in Fig. 11 rather than Fig. 9. Flow does build up from zero to a maximum with increasing density, but never falls down toward zero for higher densities. Instead, it generally falls to a value lower than the maximum flow, but still substantial, and equal to some **Saturation Flow** at the TLP. The only instances in which the throughput of a network arc goes to zero are cases of severe incidents or gridlocks at poorly managed intersections, which shall be treated as exceptions to the rule.

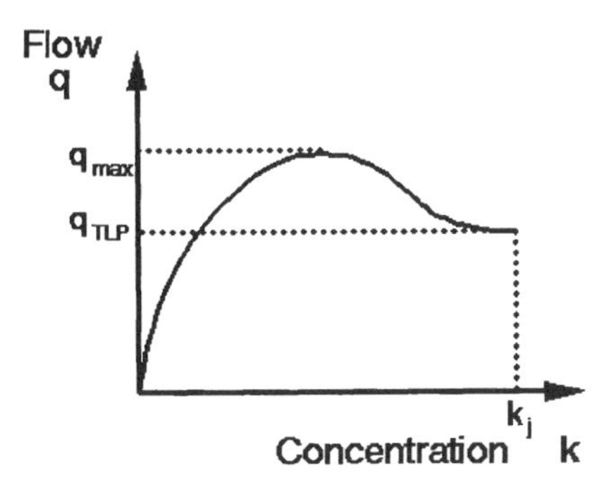

Figure 11. Modified Fundamental Diagram, with a TLP

Traffic movement along an arc is again frequently modeled today on the assumption of the existence of a Fundamental Diagram. A related modeling practice is the assumption of an **Impedance Function** for an arc, where Impedance is understood as a reduction of the capacity of the arc with increasing density. The following observations can be made regarding such modeling practices:

1) The use of an Impedance Function must be made with due consideration to the uni-directional nature of interactions between vehicles. A vehicle influences only those vehicles behind it, and not the ones in front of it. Thus, the introduction of a heavy pulse of traffic into an arc, does not alter the impedance of the whole arc instantly. If an Impedance Function is to be used, it should be limited to a relatively small region **behind** the traffic pulse.

2) Once more, it should be remembered that it is wrong to assume that the throughput of an arc goes to zero, even when vehicles are queued bumper-to-bumper along the arc. There will always be a residual throughput of an arc, equal to the throughput at its TLP.

Improved modeling of the movement of traffic along an arc during periods of congestion may be achieved as follows:

1) At low densities, the movement of individual vehicles is essentially unimpeded, with vehicles moving at their desired speed, or the speed limit. This behavior of traffic continues, up to a density approximately equal to that corresponding to the maximum throughput at the TLP.

2) When the input into the arc starts exceeding the throughput of the TLP, queueing ensues, and traffic begins to assume a stop-and-go behavior. If congestion persists to the point of completely filling the link with relatively stopped traffic, then the influence of the TLP extends to the upstream section as well, imposing the same limitation on throughput. In that domain of densities, we shall model the movement of individual traffic units as follows:

- An individual vehicle is assumed to be "served" on a First-In-First-Out (FIFO) basis. This means that, on the average, vehicles do not clear the TLP until all vehicles between the entrance to the arc and the TLP at the time of the vehicle's entrance are served. While some aggressive drivers may maneuver for improved service, the assumption of FIFO service satisfies the requirement of conservation of traffic units, and yields satisfactory results in the subsequent network optimization procedure.
- We compute a **Virtual Trajectory** for each vehicle which assumes that the vehicle moves unimpeded at constant speed, equal to its desired speed or speed limit, up to the TLP. It is stored there until all the vehicles ahead of it between the arc entrance and the TLP, at the time of the vehicle's entrance, clear the TLP.

The analytical representation of the movement of vehicles along an arc is as follows:

Let us denote by *(ij)* the arc of a network, from node i to node j. Let us then consider a vehicle entering a link ij at time t and traveling toward j with k as the next node of its travel path, (Fig. 5). Let us also assume that at time t there are $x_{ij,k}$ vehicles along the link ij with next destination k.

We estimate the departure time of the entering vehicle, from node j toward node k, on the basis of the following assumptions:

- There is a **normal travel time**, T_{ij} , from node i to node j. This time corresponds roughly to the speed limit along the link ij, and is attainable in the absence of obstructions from traffic. There is a maximum discharge rate, $u_{ij,k}$, from link *ij* toward *k*, which is achieved as long as there is a sufficient supply of vehicles in the link *ij*.

- Vehicles proceeding along *ij* toward *k* are served on a First-In-First-Out (FIFO) basis, and there is no appreciable obstruction of the traffic movement by slow vehicles. (Any such obstruction is treated as an **incident**).

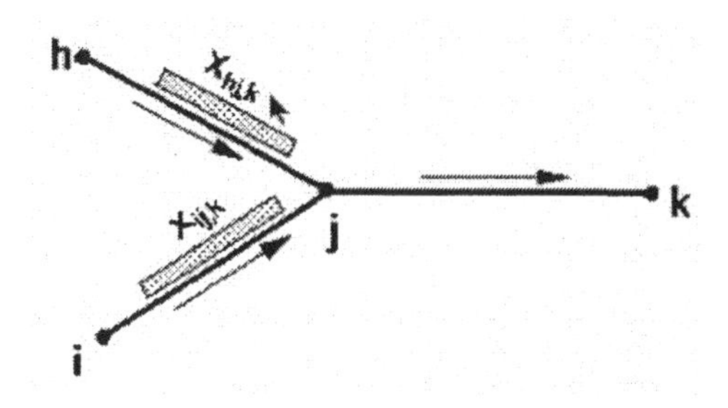

Figure 12. Queue discharge at a Throughput Limiting Point

On the basis of these assumptions, we estimate the **Virtual Trajectory** of the vehicle as follows:

- The vehicle proceeds from *i* to j in time T_{ij} .
- If

$$T_{ii} > x_{ij,k} / u_{ij,k} \tag{64}$$

the vehicle's path is not blocked. The vehicle crosses *j* and proceeds toward *k* without any delay.

- If the inequality (1) is not satisfied, the vehicle joins a **virtual queue** at *j* and waits for a period of time

$$w_{ij,k}(t) = x_{ij,k} / u_{ij,k} - T_{ij} \tag{65}$$

before being discharged toward k. The first term of the right-hand side of Eq. 65 is the time required to discharge all the vehicles which are in the link ij when the subject vehicle enters this link. Equation 65 simply states that this time is exactly equal to the time during which the entering vehicle must stay within the link.

In the preceding discussion, we have treated the node j as the TLP of the arc (ij), a reasonable assumption in most cases. The estimation of the vehicle's trajectory in effect ignores the detailed distribution of any delay incurred by the vehicle along the link ij. The vehicle is assumed to proceed from i to j at normal speed, and is stored at j in a storage "bin" of some sort until it is free to proceed toward k on a FIFO basis. Of course, care should be taken to ensure that the total storage capacity of the bin, for all traffic streams queued at j, does not exceed the physical storage limit of the roadway along ij. If it does, constraints must be imposed on the entrance of vehicles at i and, as mentioned already, the upstream link must be viewed as regulated by the same TLP.

The estimate of the travel time along different routes can now be completed by combining current information about the queues and densities at various points of the network with historical data concerning expected demand along certain routes. This may be done by an individual optimizing a route choice, or by a Traffic Management Center (TMC) aiming at the global optimization of the system. In the latter case, attainment of global optimization may be attained only if individuals follow instructions of the TMC regarding their routes.

Optimization of the network operation by the TMC requires sufficiently accurate estimates of the counts of vehicles within each link. Algorithms for the estimation of such densities have been given in Chapter 3. Once the densities along the network links are known, travel times to various destinations can be computed, and adjustments to route assignments may be undertaken, aimed at reducing the aggregate delay to all drivers. The optimal control of the network is reduced to assignment of routes to different drivers, and management of the allocation of downstream TLP throughput to converging upstream arcs. As shown by Gazis (1974), for any given assignment of routes, the allocation of TLP throughput is reduced to a parametric Linear Programming formulation. By iterating with varying choices of routes, we may then converge toward global optimization of the network.

4.5.1 Delivery of Traveller Information Services

One of the key elements of *Intelligent Transportation Systems* (ITS), which are diligently pursued around the world, is the delivery of *Advanced Traveler Information Services*, (ATIS). On the basis of the preceding discussion of Section 4.5, we can make some observations regarding the quality and content of ATIS, which can be achieved. Gazis (1995) has introduced the concept of *Service Channels* (SC) for the users of a traffic network. A Service Channel is a time and space diagram of the movement of a user's vehicle through the network such as shown in Fig. 13.

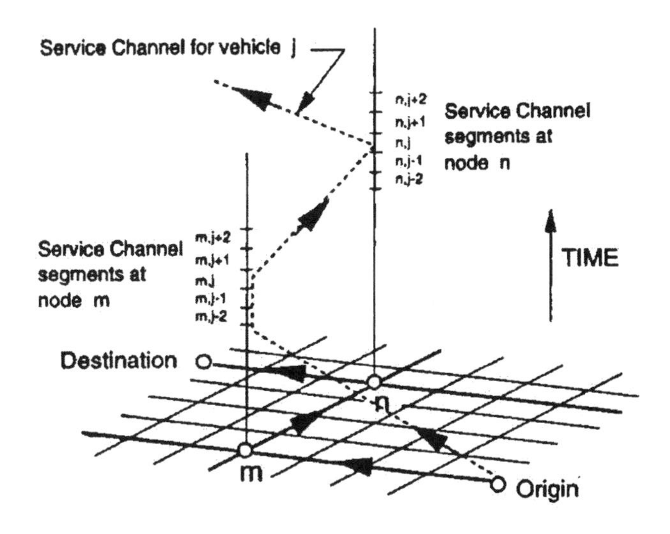

Figure 13. A Service Channel

A Service Channel is influenced by the contention between drivers for roadway space, which causes congestion, queueing and delays. Ideally, a traffic system should be operated in a way that minimizes such contention, and optimizes the system according to objectives which should include the cost of delays but other considerations such as the ability of the system to deliver services according to the desires and needs of the users.

Having the capability of estimating the travel time along any given route, and real-time adjustments to this estimate with changing traffic conditions, we may now address the question of delivery of this information to the users.

The early projections for ITS deployment intrinsically called for special instrumentation of vehicles to receive traveler information from a TMC. Early pilot projects were also associated with opinion surveys on price acceptability for such instrumentation by the driving public. As of the time of the writing of this book, The results have been less than stellar. The public does not appear to be very convinced regarding the effectiveness of ITS services in general, and does not seem willing to pay much to obtain such services.

Gazis (1998) has suggested that we might be able to accelerate deployment of ITS services by utilizing instrumentation already available, namely, the personal computers owned by many drivers and the availability of the internet. The internet has already been used to disseminate traffic information, but the dissemination has been in the form of a general description of traffic conditions in a network. In reference 4, we have promulgated the use of *Intelligent Agents* in order to provide personalized information regarding one's travel. The scenario is as follows:

An Intelligent Agent residing on a user's computer will have access to information regarding traffic conditions, both historical and real time. Before the start of a trip, the agent will estimate travel times along different routes, possibly with various starting times, and give the user pre-trip planning information on the best available option. The agent may also include modal alternatives, such a public transportation and ride sharing, in the presentation to the user. In doing so, it will try to improve the user's Service Channel without exerting any influence on the system as a whole. The activities of the Intelligent Agent can start hours before the start of the trip and be available to the user at home through a personal computer. The agent can continue monitoring the traffic conditions during the trip and transmit en route instructions directing recommended changes. The transmission can be over a telephone line, in the form of voice instructions over a cellular phone, unless the user carries a notebook in the car which can carry the Intelligent Agent's operations on board.

In the long run, Intelligent Agents such as those described above may interact with the intelligence of a TMC in order to achieve a global optimization of the operation of a network .

References

Beckman, M. J. and Golob, T. F. (1972), " Critique of Entropy and Gravity in Travel Forecasting", *Traffic Flow and Transportation*, edited by G. F. Newell, New York: Elsevier, pp. 109-117.

Bellman, R. (1958), "On a Routing Problem", *Quart. Appl. Math.*,**XVI**, 87-90.

Ben-Akiva, M. E. (1974), "Structure of Passenger Travel Demand Models", *Transportation Research Record*, **526**.

Ben-Akiva, M. E. and Lerman, S. R. (1985), *Discrete Choice Analysis: Theory and Application to Travel Demand*, Cambridge: MIT Press.

Ben-Akiva, M. E. and Bolduc, D. (1996), "Multinomial Probit with a Logit Kernel and a General Parametric Specification of the Covariance Structure", presented at the 3rd International Choice Symposium.

Ben-Akiva, M. E. and M. Bierlaire, M. (1999), "Discrete Choice Methods and Their Applications to Short Term Travel Decisions", in *Handbook of Transportation Science*, Chapter 2, Boston: Kluwer Academic Publishers.

Berman, B. R., Chinitz, B. and Hoover, E. M. (1961), *Technical Supplement to the New York Metropolitan Regional Study*, Cambrdge, MA: Harvard University Press.

Bierlaire, M., Lotan, T. and Toint, P. L. (1997), "On the Overspecification of of Multinomial and Nested Logit Models due to Alternative Specific Constants", *Transp. Sci.* **31**, 363-371.

Carey, M. (1986), "A Constraint Qualification for a Dynamic Traffic Assignment Model", *Trans. Sci.*, **20**, 55-58.

Carey, M. (1987), "Optimal Time-Varying Flows on Congested Networks", *Oper. Res.*, **35**, 58-69.

Cascetta, E. (1989). "A Stochastic Process Approach to the Analysis of Temporal Dynamics in Transportation Networks", *Transp. Res.* **23B,** 1-17.

Damm, D. (1983), "Theory and Empirical Results: A Comparison of Recent Activity-Based Research", in *Recent Advances in Travel Demand Analysis*, edited by S. Carpenter and P. Jones, London: Gower.

Dijkstra, E. W. (1959), "A Note on Two Problems in Connection with Graphs", *Numer. Math.* **1**, 269-271.

Florian, M. and Hearn, D. (1999), "Network Equilibrium and Pricing", in *Handbook of Transportation Science*, edited by R. W. Hall, Boston: Kluwer Academic Publishers, pp. 361-393.

Ford, L. R. and Fulkerson, D. R. (1962), *Flow in Networks*, Princeton: Princeton University Press.

Friesz, T. L., Bernstein, D., Smith, T. E., Tobin, M. L. and Wie, B. W. (1993), "A Variational Inequality Formulation of the Dynamic Network User Equilibrium Problem", *Oper. Res.* , **4**, 179-191 .

Gazis, D. C. (1974), "Modeling and Optimal Control of Congested Transportation Systems", *Networks*, **4,** 113-124.

Gazis, D. C. and Herman, R. (1992), "The Moving and 'Phantom' Bottlenecks", *Transp. Science*, **26,** No.3, 223-229.

Gazis, D. C. (1995), "Congestion Abatement in ITS through Centralized Route Allocation", *IVHS Journal,* **2,** 139-158.

Gazis, D. C. (1998), PASHAs: Advanced Intelligent Agents in the Service of Electronic Commerce, in *The Future of the Electronic Marketplace,* edited by D. Leebaert, MIT Press, 145-171.

Gumbel, E. J. (1958), *Statistics of Extremes*, New York: Columbia University Press.

Hansen, W. G. (1960), "Land Use Forecasting for Transportation Planning", *Highway Research Board Bulletin*, **253**, 145-151.

Helly, W. (1969), "A Time-Dependent Intervening Opportunity Land Use Model", *Socio-Economic Planning Sci.*, **3**, 65-73.

Helly, W. (1974), "Traffic Generation, Distribution, and Assignment", *Traffic Science*, edited by D. C. Gazis, New York: John Wiley & Sons, Chapter 4.

Hirsh, M, Prashkea, J. N. and Ben-Akiva, M. (1986), "Dynamic Model of Weekly Activity Pattern", *Transp. Sci.*, **20**, 24-36.

Ho, J. K. (1980), "A Successive Linear Optimization Approach to the Dynamic Traffic Assignment Problem", *Trans. Sci.*, **14**, 295-305.

Jones, P. (1983), "The Practical Application of Activity Based Approaches in Transport Planning: An Assessment", in *Recent Advances in Travel Demand Analysis*, edited by S. Carpenter and P. Jones, London: Gower.

Luce, R. (1959), *Individual Choice Behavior: A Theoretical Analysis*, New York: J. Wiley and Sons.

McFadden, D. (1978), "Modeling the Choice of Residual Location", in *Spatial Interaction Theory and Residential Location*, Amsterdam: North Holland, pp. 75-96.

McFadden, D. (1989), "A Method of Simulated Moments for Estimation of Discrete Response Models Without Numerical Integration", *Econometrica*, **57**, 995-1026.

Merchant, D. K. (1974), *A Study of Dynamic Traffic Assignment and Control*, PhD Dissertation. Cornell Univ.

Merchant, D. K. and Nemhauser, G. L. (1978a), "A Model and an Algorithm for the Dynamic Traffic Assignment Problem", *Trans. Sci.*, **12**, 163-199.

Merchant, D. K. and Nemhauser, G. L. (1978b), "Optimality Conditions for a Dynamic Traffic Assignment Problem"" *Trans. Sci.*, **12**, 200-207.

Murchand, J. D. (1969), "Bibliography of the Shortest Route Problem", *London Business Studies Report*, **LBS-TNT- 6.2**.

Pas, E. I. (1982), "Analytically-Derived Classification of Daily Travel-Activity Behavior: Description, Evaluation and Interpolation", Transp. Res. Board.

Potts, R. B. and Oliver, R. M. (1972), *Flows in Transportation Networks*, New York: Academic.

Putman, S. H. (1966), "Analytic Models fr Implementing the Economic Impact Studies for the Northeast Corridor Transportation Project", 30th National Meeting, Operations Research Society of America.

Ruiter, E. R. (1967). "Improvements in Understanding, Calibrating, and Applying the Opportunity Model", *Highway Res. Rec.*, **165**, 1-21.

Schneider, M. (1960), *Chicago Area Transportation Study*, Vol. 2, 1960.

Sherratt, G. G. (1960), "A Model for General Urban Growth", *Management Sciences – Models and Techniques*, New York: Pergamon, Vol. 2, pp.147-159.

Stopher, P. R., Meyburg, A. H. and Brog, W., editors, (1979), *Proc.of the 4th Intern. Conf. On Behavioral Travel Modeling*, Grainau, Germany.

Stouffer, S. A. (1940), "Intervening Opportunities", *Amer. Sociological Rev.*, **5**, 845-867.

Swerdloff, C. N. and Stowers, J. R. (1966), "A Test of Some First Generation Residential Land Use Models", 45th Highway Research Board Meeting, Washington, DC.

Tversky, A. (1972), "Elimination by Aspects: A Theory of Choice", *Psychological Review*, **79**, 281-299.

Voorhees, A. M. (1955), "A General Theory of Traffic Movement", *Proc.Inst.Traffic Engrs*, 46-56.

Wardrop, J. G. (1952), "Some Theoretical Aspects of Road Traffic Research", *Proc. Inst. Civil Engineers*, Part II, 325-378.

Wilson, A. G. (1967), "A Statistical Theory of Spatial Distribution Models", *Transport. Res.*, **1**, 253-269.

Yagar, S. (1970), *Analysis of the Peak Period Travel in a Freeway-Arterial Network*, PhD Dissertation, Univ. of California, Berkeley.

Yagar, S. (1971), "Dynamic Traffic Assignment by Individual Path Minimization and Queueing", Trans. Res., **5**, 170-196.

Wie, B. W. (1991), "Traffic Assignment, Dynamic", in *Concise Encyclopedia of Traffic & Transportation Systems*, edited by M. Papageorgiou, Oxford: Pergamon Press, pp. 521-524.

Wu, J. H., Chen, Y. and Florian, M. (1997a), "The Continuous Dynamic Network Loading Problem: A Mathematical Formulation and Solution Method", *Transp. Res.* **32B**, 173-187.

Xu, Y., Wu, J. H., Florian, M., Marcotte, P. and Zhu, D. L. (2002), "Advances in the Continuous Dynamic Network Loading Problem", *Transp. Sci.*, (in press).

Additional references

Aashtiani, H. Z. and Magnanti, T. L. (1981), "Equilibria in a Congested Transportation Network", *SIAM J. of Algebraic and Discrete Methods*, **2**, 213-226.

Arezki, Y. and Van Vliet, D. (1990), "A Full Analytical Implementation of the PARTAN/Frank-Wolfe Algorithm for Equilibrium Assignment", *Transp. Sci.*, **24**, 58-62.

Beckmann, M. J. and Golob, T. F. (1972), "A Critique of Entropy and Gravity in Travel Forecasting", *Traffic Flow and Transportation*, edited by G. F. Newell, New York: Elsevier, pp.109-117.

Dafermos, S. (1972), "The Traffic Assignment Model forMulti-Class Transportation Network", *Transp. Sci.*, **6**, 73-87.

Dafermos, S. and A. Nagurney, A. (1984), "Sensitivity Analysis for the Asymmetric Network Equilibrium Problem", *Math. Programming*, **28**, 174-184.

Dow, P. and Van Vliet, D. (1979), "Capacity Restrained Road Assignment", *Traffic Eng. and Control*, 261-273.

Evans, S. P. (1976), "Derivation and Analysis of Some Models for Combining Trip Distribution and Assignment", *Transp. Res.*, **10**, 37-57.

Florian, M. and Nguyen, S. (1976), "An Application and Validation of Equilibrium Trip Assignment Methods", *Transp. Sci.*, **10**, 379-389.

Florian, M. and Spiess, H. (1983), "On Binary Mode Choice/Assignment Models", *Transp. Sci.*, **17**, 32-47.

Florian, M., Guelat, Y. and Spiess, H. (1987), "An Efficient Implementation of the PARTAN variant of the Linear Approximation for the Network Equilibrium Problem", *Networks*, **17**, 319-339.

Florian, M. and Hearn, D. (1995), "Network Equilibrium Models and Algorithms", Chapter 6 in *Handbook in OR and MS, Vol. 8*, M. Ball et al, editors, pp.485-550.

Fratar, T. J. (1954), "Forecasting Distribution of Interzonal Vehicular Trips by Successive Approximations", *Proc. Highway Res. Brd.*, **33**, 376-385.

Friedrich, P. (1956), "Die Variationsrechnung als Planungsverfahren der Stadt und Landesplanung", *Schriften der Akademie fur Raumforschung und Landesplanung*, Brenen-Horn: Walter Dorn, Vol. 32.

Hall, M. (1978), "Properties of the Equilibrium State in Transportation Networks", *Transp. Sci.*, **12**, 208-216.

Hamburg, J. R. and Sharkey, R. H. (1961) "Land Use Forecast", *Chicago Area Transportation Study*, Chicago, 3.2.6.10.

Harmann, F. V. (1964), *Population Forecasting Methods*, Washington, DC: Bureau of Public Roads.

Harris, B. (1962), "Linear Programming and the Projection of Land Users", *Penn–Jersey Transportation Study*, **PJ** paper No. 20.

Helly, W. (1965), "Efficiency in Road Traffic Flow", *Proc. 2nd Intern, Symp. on Theory of Traffic Flow*, Paris: OECD, pp.264-275.

Herbert, J. D. and B. H. Stevens, B. H. (1960), "A Model for the Distribution of Residential Activity in Urban reas", *J. Regional Sci. Assoc.*

Hoshino, T. (1972), "Theory of Traffic Assignment to a Road Network", *Traffic Flow and Transportation*, edited by G. F. Newell, New York: Elsevier, pp.195-214.

Knight, F. H. (1981), "Some Fallacies in the Interpretation of Social Costs", *Quarterly J. of Economics*, **38**, 306-312.

Lan, T. and Newell, G. F. (1967), "Flow Dependent Traffic Assignment in a Circular City", *Transp. Sci.*, **1**, No. 4, 318-361.

Leblanc, L. J., Morlok, E. K. and Pierskalla, W. P. (1975), "An Efficient Approach to Solving the Road Network Equilibrium Traffic Assignment Problem", *Transp. Sci.*, **5**, 309-318.

Nguyen, S. (1976), "A Unified Approach to Equilibrium Methods for Traffic Assignment", in *Traffic Equilibrium Methods*, Proceedings 1974, edited by M. Florian, vol. **118**. Lecture notes in Economics and Mathematical Systems, New York: Springer Verlag, pp. 148-182.

Papageorgiou, M. (1990), "Dynamic Modeling, Assignment, and Route Guidance in Traffic Networks", *Transp. Res.*, **24B**, 471-495.

Patriksson, P. (1983), *The Traffic Assignment Problem: Models and Methods*, VNU Science Press, p. 223.

Powell, W. B. and Sheffi, Y. (1982), "The Convergence of Equilibrium Algorithms with Predetermined Step Sizes", *Transp. Sci.*, **16**, 45-55.

Sheffi, Y. and Powell, W. B. (1982), " An Algorithm for the Equilibrium Assignment Problem with Random Link Times", *Networks*, **12**, 191-207.

Smeed, R. J. (1961), *The Traffic Problem in Towns*, Manchester: Manchester Statistical Society.

Smith, M. J. (1979), "Existence, Uniqueness and Stability of Traffic Equilibria", *Transp. Res.*, **13B**, 295-304.

Smith, M. J. (1993), "A New Dynamic Traffic Model and the Existence and Calculation of Dynamic User Equilibria On Congested Capacity Constrained Road Network", *Transp. Res.*, **27B**, 49-63.

Swerdloff, C. N. and Stowers, J. R. (1966), "A Test of Some First Generation Residential Land Use Models", *5th Highway Res. Brd. Meeting*, Washington, DC.

U. S. Department of Commerce, Bureau of Public Roads, *Traffic Assignment Manual*, Washington, DC, 1964.

U. S. Department of Commerce, Bureau of Public Roads, *Highway Capacity Manual*, Washington, DC, 1965.

Voorhees, A. M. (1955), "A General Theory of Traffic Movement", *Proc. Inst. Traffic Engrs.*, 46-56.

Wie, B., Tobin, R., Bernstein, R. and Friesz, T. (1995), "Comparison of System Optimum and User Equilibrium Dynamic Traffic Assignment with Schedule Delays", *Transp. Res.*, **3C**, 389-411.

Appendix

Application of Kalman Filtering for Density Estimation in Traffic Networks

We shall outline here the application of the methodology of the *Discrete-Time, Extended Kalman Filtering* for estimating the densities in various links of a traffic network. The application of this methodology is useful in systems where surveillance is limited to detection of passage of vehicles over detectors placed at discrete points along the roadway, for example, at the entrance and exit points of a network link. Assuming that one can initialize the state of the system through knowledge of the count of vehicles in each section at some point in time, the counts may be updated subsequently by adding the number of vehicles entering a section and subtracting the number of vehicles exiting. However, most detectors are subject to errors in measuring the exact number of vehicles passing over them, and the cumulative error may cause a substantial deviation from the real count. The Kalman Filtering methodology may be applied in order to substantially reduce such errors. In particular, the "Discrete-Time, Extended Kalman Filter[1]" has been proposed by Szeto and Gazis[2] for estimating densities in a "discrete-time control system", such as a traffic network, which is monitored and controlled at regular time intervals.

The Extended Kalman Filter is particularly appropriate for estimating the state variables in a nonlinear (noisy) system with nonlinear (noisy) observations. The algorithm re-linearizes the dynamics about each new state variable estimate, thus not allowing estimation errors to propagate through time. It does so by combining two essentially independent estimates of the state variables, one obtained from the "state equation", and another one obtained from an "observation equation", both subject to errors which are uncorrelated. The resulting estimates correspond to a minimum value of the integral of the squared deviation from true values of the state variables.

Let $k = 0, 1, 2, \ldots$. Denote a time index, where $k = 0$ corresponds to a starting time. Consider a discrete-time, nonlinear, and possibly time-varying system, whose state vector x_k is an n-dimensional vector $\underline{x}_k \in R_n$. Assume that the state changes according to the stochastic difference equation

$$x_{k+1} = f_k(\underline{x}_k) + \xi_k \quad , \tag{1}$$

where ξ_k is the error (noise). Let us suppose that independent observations provide a measurement vector z_k ,$(z_k \in R_r)$ which is nonlinearly related to the state vector, and has the form

$$z_k = h_k(x_k) + \theta_k \ , \tag{2}$$

where θ_k is also a noise term. Let us assume that the state non-linearity,

$$f_k(\cdot) : R_n \to R_n \ , \tag{3}$$

and the output non-linearity,

$$h_k(\cdot) : R_n \to R_r \ , \tag{4}$$

are known, continuous, and sufficiently differentiable. Also assume that the initial state vector x_0 is an n-dimensional random variable with known mean

$$\bar{x}_0 = E\{x_0\} , \tag{5}$$

and covariant matrix

$$\Sigma_0 = \mathrm{cov}[x_0 ; x_0] = E\left\{ \left(x_0 - \bar{x}_0 \right)\left(x_0 - \bar{x}_0 \right)' \right\} , \tag{6}$$

where primes denote vector transposition. Now, assume zero-mean, uncorrelated noise in both the state and the observation equations, leading to

$$\begin{aligned}
&E\{\xi_k\} = 0 , \\
&E\{\theta_k\} = 0 , \\
&\mathrm{cov}\{\xi_k ; \xi_j\} = E\{\xi_k \xi_j'\} = \Xi_k \delta_{kj} , \\
&\Xi_k = \Xi_k' \geq 0 , \\
&\mathrm{cov}\{\theta_k ; \theta_j\} = E\{\theta_k \theta_j'\} = \Theta_k \delta_{kj} , \\
&\Theta_k = \Theta_k' > 0 ,
\end{aligned} \tag{7}$$

for all k. Furthermore, assume that x_0 , ξ_k , θ_j are mutually independent for all k , j.

We shall use the following notation:

$\hat{x}_{k|k}$ = estimate of the state vector x_k based on the observations of the variables $z_1, z_2, \cdots, z_k$,

$\hat{x}_{k+1|k}$ = predicted estimate of the vector x_{k+1} based only on the observations $z_1, z_2, \cdots, z_k$, (i.e., before the measurement z_{k+1} is made) ,

$$\Sigma_{k|k} = E\left\{\left(x_k - \hat{x}_{k|k}\right)\left(x_k - \hat{x}_{k|k}\right)'\right\} ,$$

$$\Sigma_{k+1|k} = E\left\{\left(x_{k+1} - \hat{x}_{k+1|k}\right)\left(x_{k+1} - \hat{x}_{k+1|k}\right)'\right\} .$$

If the error vectors have zero mean, then the second moment matrices, Σ, are the covariant matrices of the state estimation error. Now, the discrete-time extended Kalman Filter algorithm is best explained by decomposing it into three different steps:

(a) *Initialize* :

Start the algorithm by setting

$$\hat{x}_{0|0} = \bar{x}_0 ,$$
$$\Sigma_{0|0} = \Sigma_0 . \tag{8}$$

(b) *Predict* :

Generate $\hat{x}_{k+1|k}, \Sigma_{k+1|k}$ according to

$$\hat{x}_{k+1|k} = f_k\left(\hat{x}_{k|k}\right),$$
$$\Sigma_{k+1|k} = \hat{F}_k \Sigma_{k|k} \hat{F}_{k'} + \Xi_k , \tag{9}$$

where

$$\hat{F}_k = \frac{\partial f_k\left(x_k\right)}{\partial x_k} \quad \text{for} \quad x_k = \hat{x}_{k|k} . \tag{10}$$

(c) *Update*:

Generate $\hat{x}_{k+1|k+1}, \Sigma_{k+1|k+1}$ according to

$$\hat{x}_{k+1|k+1} = \hat{x}_{k+1|k} + G_{k+1}\left\{z_{k+1} - h_{k+1}\hat{x}_{k+1|k}\right\},$$
$$\Sigma_{k+1|k+1} = \Sigma_{k+1|k} - G_{k+1}\hat{H}_{k+1}\Sigma_{k+1|k} \,, \tag{11}$$

where G_{k+1} is known as the "Kalman gain matrix", and is given by

$$G_{k+1} = \Sigma_{k+1|k}H'_{k+1}\left\{H_{k+1}\Sigma_{k+1|k}H'_{k+1} + \theta_{k+1}\right\}^{-1} \,,$$
$$H_{k+1} = \frac{\partial h_{k+1}(x_{k+1})}{\partial x_{k+1}} \quad for \;\; x_{k+1} = \hat{x}_{k+1|k} \,. \tag{12}$$

The algorithm is recursive in nature. Starting even with an initial guess at the state variables, it generates a new state estimate each time an observation vector becomes available, moving towards constant improvement of the estimates.

Let us consider a network link shown in Fig. 1, with length L, and sensors at the entrance and exit capable of recording the number of vehicles passing over them and the speed associated with each vehicle. Knowing that the count measurements are subject to error, we wish to use the speed and count measurements in order to generate, at discrete intervals, improved estimates of the number of vehicles within the link.

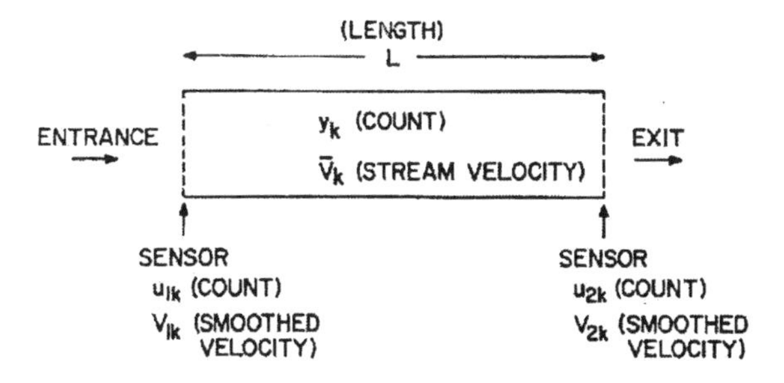

Figure 1. A sensor-based surveillance of a network link

Let y_k denote the number of vehicles in the link at the kth discrete-time interval, u_{1k} denote the number of vehicles entering the link between time k and $k+1$, and u_{2k} the number of vehicles exiting the link in the same time interval. Assuming that the input and output sensor measurements are subject to error, we have the following relationship:

$$y_{k+1} = y_k + u_{1k} - u_{2k} + \xi_k \ , \tag{13}$$

where ξ_k is the error (noise) term. Equation 13 is the "state equation" in the Kalman Filter nomenclature. We now wish to utilize the speed information in order to produce another equation, the "observation equation". In order to do this, we have to translate the speed information into one related to the count of vehicles in the link, or the traffic density therein.

Let V_{1k} be the exponentially smoothed velocity at time k, at the entrance, and V_{2k} the exponentially smoothed velocity at the exit of the link. Thus, V_{1k} and V_{2k} are estimates of the average speed of vehicles entering and exiting the link, respectively. The space mean speed of vehicles within the section at time k, $\overline{V}_k$, may be approximated by an expression of the form

$$\overline{V}_k = \alpha V_{1k} + (1-\alpha)V_{2k} \ , \tag{14}$$

where α is a weighting factor. Equation 14 implies that vehicles within the link attain their velocity through a combination of forward and backward propagation of speed changes. For moderately heavy traffic, but relatively uncongested conditions, one may expect a value of $\alpha > 0.5$, say, between 0.5 and 0.7, because of the one-sided nature of traffic flow. It is possible to dynamically modify the value of α, in order to take into account the backward and forward propagation of speed changes in a real application.

The space mean speed V_k is a variable that measures the mean speed of the entire stream of traffic. In Chapter 1 of this book, we discussed various models relating such speed measurements to traffic densities. Although the models discussed there are essentially appropriate for an infinitely long roadway, they may be used for obtaining an approximation to the link density from speed measurements. In general, the stream velocity decreases with increasing density. For the Lincoln tunnel experiment discussed in Chapter 3, we found that a reasonable relationship between speed and density was

$$\overline{V}_k \cong b\exp\left\{-\frac{1}{2}[(y_k/L)/a]^2\right\} , \tag{15}$$

where L is the link length and a and b are constant parameters. When the stream velocity V_k is plotted against the density y_k/L, we obtain a bell-shaped curve such as that shown in Fig. 2. Again the deviations from the approximation of Eq. 15 can be accommodated by introducing a noise term, leading to the "observation rquation"

$$\overline{V}_k \cong b\exp\left\{-\frac{1}{2}[(y_k/L)/a]^2\right\} + \theta_k , \tag{16}$$

where θ_k is the noise term. The parameters a and b may be regarded as time-invariant variables defined for each observation period, and consequently satisfying the relationships

$$\begin{aligned} a_{k+1} &= a_k , \\ b_{k+1} &= b_k . \end{aligned} \tag{17}$$

The parameters a_k and b_k may be viewed as additional state variables of the system.

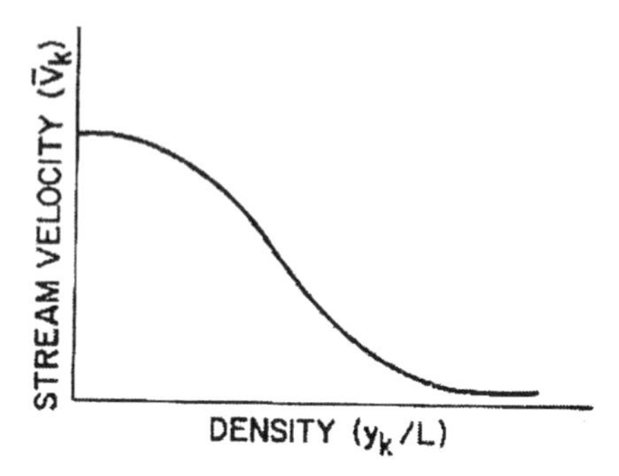

Figure 2. Speed-density relationship

We now observe that Eqs. 13 and 17 form a set of "state equations" of the form of Eq. 1, and Eq. 16 is an "observation equation" of the form of Eq. 2. The state variables are the nuber of vehicles in the link, y_k , and the parameters a_k and b_k. The observed variable is the stream velocity V_k, which is generated from the sensor measurements according to Eq. 14, and leading to an "observed" value for the density, or count, of vehicles within the section, which can be obtained from Eq. 16. The system is driven by the counts of vehicles entering and exiting the link, plus the noise in both the state and the observation equations. We can now obtain the state variable estimates by using the extended Kalman Filter described above. The computations are lengthy but straightforward, and they are not given here.

The application of the Kalman Filter was tested against the data obtained from the Lincoln Tunnel experiment[3] described in Chapter 3. As indicated in that discussion, very accurate values for the densities were available for each section of the tunnel through identification of individual vehicles or combinations of vehicles as they passed over successive sensors. These accurate values were used in order to test the power of the Kalman Filtering approach, which was found to be more than adequate for most situations. Figure 3 shows a comparison of actual vehicle count data, for a single tunnel section, with counts estimated by Kalman Filtering.

The effectiveness of the Kalman Filter approach in estimating traffic counts and densities depends strongly on the reliability of the "observation" data. In the preceding discussion, the observation data were obtained on the basis of a phenomenological relationship between speed and density, and a simple averaging of speed data at the two ends of the link. Such a procedure is likely to succeed as long as traffic is moving along the link, but it is likely to fail when congestion sets in and a number of cars are "parked" in a queue and waiting for service. In such a case, neither the averaging of the speed nor the phenomenological relationship are reliable. However, one may seek alternative "observation" data for the Kalman Filter. As an example, let us consider the case where there is some partial identification of some vehicles, providing a handle for estimating the count of vehicles within the link.

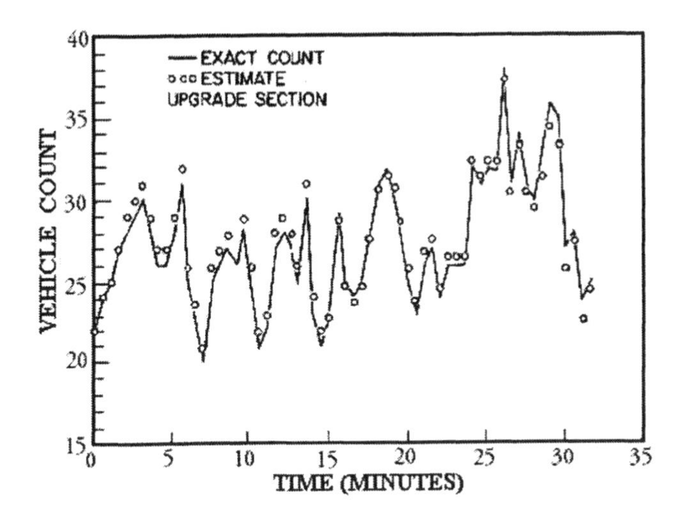

Figure 3. Comparison of exact counts and Kalman Filtering estimates.

The identification of some vehicles may be obtainable through the use of RFID (Radio Frequency ID) devices, such as those used for electronic toll payment. The RFID devices can identify individual vehicles at the entrance and exit of a ramp if we simply install receiving antennas at those points. These "tagged" vehicles can then be used as markers, whose passage over the entrance and exit detectors define a time segment during which the number of vehicles entering the link can be measured fairly accurately. To be sure, we now in general will have to deal with multi-lane rather than single lane traffic. However, we can use the information from the tagged vehicles for estimating densities in the following ways.

1) At low densities, we can rely on the standard Kalman filtering methodology, using speed-flow relationships for a second estimate of the density. Alternatively, we can use the following computation. Let us assume that the tagged vehicle covered the distance from the origin i to the

destination j of the link in time t_{ij} , (Fig. 4). This corresponds to an average travel speed v_{ij} given by

$$v_{ij} = L_{ij} / t_{ij} \tag{18}$$

where L_{ij} is the length of the link.

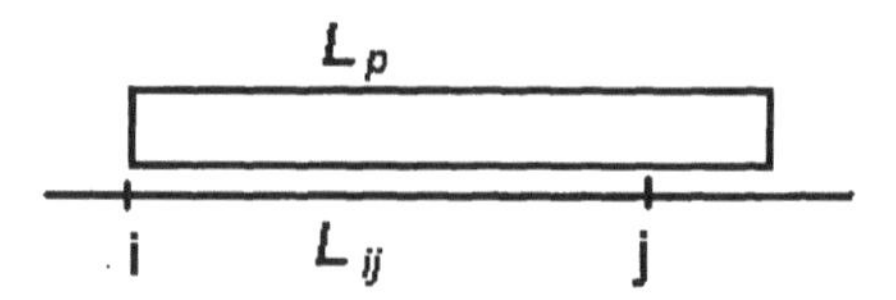

Figure 6. Travel across an arc ij.

During the time of the passage of the tagged vehicle, we measure the entrance speeds of all the entering vehicles and compute their average, v_p . This defines a "spread" of the aggregate of these vehicles equal to

$$L_p = v_p \, . \, t_{ij} \tag{19}$$

The quantity L_p will be larger or smaller than L_{ij} , depending on whether or not the tagged vehicle is slower or faster than the platoon entering behind it. It is plotted as larger in Fig. 6, implying that the tagged vehicle is slower. We can then use as a second estimate for the arc density the quantity

$$k_{ij} = N_p / L_p \tag{20}$$

where N_p is the number of vehicles in the platoon. It may be pointed out that in the case of low densities, the moving average of the travel time across an arc of several vehicles may be used in computing travel times along different routes of the network.

2) As the density increases and congestion sets in, one may assume that passing becomes relatively difficult and any tagged vehicle moves together with the rest of the vehicles. In this case, the travel time of any tagged vehicle is an accurate measure of the travel time of most of the vehicles entering the link. We can simply count all the vehicles entering the link from the time a tagged vehicle enters to the time it leaves the link, and use this count as the observation count.

The preceding discussions treated traffic links as isolated ones, and therefore any errors in measurement of the passage of cars over entrance and exit sensors were assumed as independently generated. In fact, links are generally in tandem, and errors of connected links are correlated, since the output error for an upstream link will be the input error of the downstream link. This observation may be used to further improve the density estimation. The Kalman filtering methodology may be revised to include the correlation of errors for arcs in tandem, and adjustments to the final estimates of densities can be made on arguments of correlation of errors. An improved Kalman Filtering approach utilizing this correlation of errors has been proposed by Gazis and Liu[4]. Preliminary results, based on actual and computed values of traffic counts for Freeway 41, in Fresno, California, indicate a substantial improvement of the accuracy of count estimates when the correlation of errors is explicitly considered.

References

1. Gazis, D. C. and Foote, R. S., 1969, "Surveillance and Control of Tunnel Traffic by an On-Line Digital Computer", *Trans. Sci.*, **3**, 255-275.
2. Jazwinski, A. H., 1970, *Stochastic Processes and Filtering Theory,* New York: Academic Press.
3. Szeto, M. W. and Gazis, D. C., 1972, "Application of Kalman Filtering to the Surveillance and Control of Traffic Systems" , *Trans. Sci.* , **6**, 419-439.

Index